Environmental Humanities of Extraction in Africa

This book brings together perspectives on resource exploitation to expose the continued environmental and socio-political concerns in post-colonial Africa. The continent is host to a myriad of environmental issues, largely resulting from its rich diversity in natural resources that have been historically subjected to exploitation.

Colonial patterns of resource use and capital accumulation continue unabated, making environmental and related socio-political problems a dominant feature of African economies. The book pursues the manifestation of these problems through four themes: environmental justice, violent capitalocenes, indigenous knowledge, and climate change. The editors locate the book within the broad fields of political ecology and environmental geopolitics to highlight the intricate geographies of resource exploitation across Africa. It uniquely focuses on the socio-political and geopolitical dynamics associated with the exploitation of Africa's natural resources and its people. The case studies from different parts of Africa tell a compelling story of resource exploitation, related issues of environmental degradation in a continent particularly vulnerable to climate change, and the continued plundering of its natural resources.

The book will be of great interest to scholars and students from the interdisciplinary fields of the environmental humanities and environmental studies more broadly, as well as those studying political ecology, environmental policy, and natural resources with a specific focus on Africa.

James Ogude is Director of the Centre for the Advancement of Scholarship, University of Pretoria, South Africa.

Tafadzwa Mushonga is a research fellow at the Centre for the Advancement of Scholarship, University of Pretoria, South Africa.

Routledge Environmental Humanities
Series editors: Scott Slovic (University of Idaho, USA),
Joni Adamson (Arizona State University, USA) and
Yuki Masami (Aoyama Gakuin University, Japan)

Editorial Board

Christina Alt, St Andrews University, UK

Alison Bashford, University of New South Wales, Australia

Peter Coates, University of Bristol, UK

Thom van Dooren, University of Sydney, Australia

Georgina Endfield, Liverpool, UK

Jodi Frawley, University of Western Australia, Australia

Andrea Gaynor, The University of Western Australia, Australia

Christina Gerhardt, University of Hawai'i at Mānoa, USA

Tom Lynch, University of Nebraska, Lincoln, USA

Iain McCalman, Australian Catholic University, Australia

Jennifer Newell, Australian Museum, Sydney, Australia

Simon Pooley, Imperial College London, UK

Sandra Swart, Stellenbosch University, South Africa

Ann Waltner, University of Minnesota, US

Jessica Weir, University of Western Sydney, Australia

International Advisory Board

William Beinart, University of Oxford, UK

Jane Carruthers, University of South Africa, Pretoria, South Africa

Dipesh Chakrabarty, University of Chicago, USA

Paul Holm, Trinity College, Dublin, Republic of Ireland

Shen Hou, Renmin University of China, Beijing, China

Rob Nixon, Princeton University, Princeton NJ, USA

Pauline Phemister, Institute of Advanced Studies in the Humanities, University of Edinburgh, UK

Sverker Sorlin, KTH Environmental Humanities Laboratory, Royal Institute of Technology, Stockholm, Sweden

Helmuth Trischler, Deutsches Museum, Munich and Co-Director, Rachel Carson Centre, Ludwig-Maximilians-Universität, Germany

Mary Evelyn Tucker, Yale University, USA

Kirsten Wehner, University of London, UK

The *Routledge Environmental Humanities* series is an original and inspiring venture recognising that today's world agricultural and water crises, ocean pollution and resource depletion, global warming from greenhouse gases, urban sprawl, overpopulation, food insecurity and environmental justice are all *crises of culture*.

The reality of understanding and finding adaptive solutions to our present and future environmental challenges has shifted the epicenter of environmental studies away from an exclusively scientific and technological framework to one that depends on the human-focused disciplines and ideas of the humanities and allied social sciences.

We thus welcome book proposals from all humanities and social sciences disciplines for an inclusive and interdisciplinary series. We favour manuscripts aimed at an international readership and written in a lively and accessible style. The readership comprises scholars and students from the humanities and social sciences and thoughtful readers concerned about the human dimensions of environmental change.

For more information about this series, please visit: www.routledge.com/Routledge-Environmental-Humanities/book-series/REH

Environmental Humanities of Extraction in Africa

Poetics and Politics of Exploitation

Edited by James Ogude and
Tafadzwa Mushonga

First published 2023
by Routledge
4 Park Square, Milton Park, Abingdon, Oxon OX14 4RN

and by Routledge
605 Third Avenue, New York, NY 10158

Routledge is an imprint of the Taylor & Francis Group, an informa business

© 2023 selection and editorial matter, James Ogude and Tafadzwa Mushonga; individual chapters, the contributors

The right of James Ogude and Tafadzwa Mushonga to be identified as the authors of the editorial material, and of the authors for their individual chapters, has been asserted in accordance with sections 77 and 78 of the Copyright, Designs and Patents Act 1988.

All rights reserved. No part of this book may be reprinted or reproduced or utilised in any form or by any electronic, mechanical, or other means, now known or hereafter invented, including photocopying and recording, or in any information storage or retrieval system, without permission in writing from the publishers.

Trademark notice: Product or corporate names may be trademarks or registered trademarks, and are used only for identification and explanation without intent to infringe.

British Library Cataloguing-in-Publication Data
A catalogue record for this book is available from the British Library

Library of Congress Cataloging-in-Publication Data
A catalog record for this book has been requested

ISBN: 978-1-032-26360-1 (hbk)
ISBN: 978-1-032-26361-8 (pbk)
ISBN: 978-1-003-28793-3 (ebk)

DOI: 10.4324/9781003287933

Typeset in Sabon
by Apex CoVantage, LLC

Contents

List of Figures	ix
List of Tables	x
List of Contributors	xi
List of Abbreviations	xv

1. Introduction: The Intractable Problem: Africa and the Pitfalls of Resource Exploitation in a Globalising World   1
   TAFADZWA MUSHONGA AND JAMES OGUDE

2. Petitioning the Future Through Environmental Justice. A Reading From Angola   23
   CATARINA ANTUNES GOMES

3. African Goats, the State and Conservation in Colonial Zimbabwe, 1892–1970s   38
   ELIJAH DORO

4. The Politics of Exclusion and Violence in Protected Areas   56
   TAFADZWA MUSHONGA

5. The Politics of Mining Pollution in Zambia: Investigating 100 Years of Environmental Management on the Copperbelt   75
   CHIBAMBA JENNIFER CHANSA

6. "Aesthetics of the Earth": African Literature as a Witness to Post-colonial Ecology   93
   JAMES OGUDE

7. "One in Heart as They Are in Tongue": "Yoruba," Land and Environmental Violence in Colonial Southwestern Nigeria   105
   AYODEJI WAKIL ADEGBITE

8 A Revaluation of Traditional Ecological Thoughts, Knowledge, and Practices of the Aari of Southern Ethiopia 123
ENDALKACHEW HAILU GULUMA AND ZEWDE JAGRE DANTAMO

9 Climate Injustice: How It Is Affecting Africa 139
D. A. MASOLO

10 International Environment Law, the Humanities Nexus, and Some Reflections on *Creative Legal Solutions* 153
GEORGE ODERA OUTA

11 Carbon Dioxide, Climate Change, and an Energy Transition for a Future Africa 170
EMIL RODUNER AND EGMONT R. ROHWER

12 Poetics and Politics of Resource Exploitation in Africa: Insights From Chapters 179
JAMES OGUDE AND TAFADZWA MUSHONGA

Index 184

Figures

2.1	Cacimba (handmade community non-drinkable water source for human and animal consumption and agriculture), Tchiquaqueia. Huíla	31
3.1	Impound notes of stray African goats	43
4.1	Converging actors, converging interests in Sikumi Forest Reserve, Zimbabwe	60
4.2	Structural and physical barriers in resource extraction	66
8.1	Map of Aariland (areas indicated as Semen Ari and Debub Ari)	125
11.1	South African mining map with an indication for the Cradle of Humankind	171
11.2	Satellite view of the Earth at night	175
11.3	Africa's rich and precious ethnic diversity, overlaid with country borders	176

Tables

2.1 Chronic undernutrition by Angolan Provinces – children under 5 years old (IIMS 2015–2016). 30
3.1 Statistics of goats in Southern Rhodesia, 1911–1917. 46
3.2 Commercial slaughter of African goats in Southern Rhodesia, 1968–1973. 50

Contributors

Ayodeji Adegbite is a PhD candidate in the Department of History at the University of Wisconsin–Madison. His research focuses on the history of (global)health, infectious diseases, and environmental and social change in Africa. Ayodeji received his bachelor's degree in History and International Studies from the University of Ilorin, Nigeria, in 2014. He also acquired a master's degree in Peace and Development Studies at the Center for Peace and Strategic Studies, Unilorin. At UW–Madison, he received a master's degree in History of Science, Medicine, and Technology. He has taught Global Environmental Health at the University of Wisconsin–Madison.

Chibamba Jennifer Chansa is a Postdoctoral Fellow at the University of the Free State in the International Studies Group. She holds a BA in History and Library and Information Studies from the University of Zambia, MA in African Studies from the University of Basel, and PhD in Africa Studies from the University of the Free State. Jennifer's research interests include environmental, mining, and labour history; as well as African anthropology. Her MA research focused on the "urbanisation" of colonial mine labour in Zambia, drawing from the fieldwork of British social anthropologist Godfrey Wilson. Jennifer's PhD dissertation examined the environmental pollution and management within the "old" (Copperbelt) and "new" (North-Western) Copperbelt mining regions of Zambia since independence. She is currently working towards expanding her PhD research, particularly on the "new" Copperbelt. Jennifer has previously taught in the University of Zambia's History Department.

Zewde Jagre Dantamo (M.A.) is currently an Assistant Professor of Literature in the Department of English Language and Literature, Arba Minch University. He teaches English Language and Literature courses and is presently working on three research projects. The first is titled, *A Structuralist Study of Brale Folk Narratives, Gender in Gamo Language*. The second is titled *Orature: A Critical Discourse Analysis*. The third is titled *Collection Classification and Documentation of Koore Oral Narratives*.

Elijah Doro is a postdoctoral research fellow at the Centre for Advancement of Scholarship, University of Pretoria. His research is on environmental histories of southern Africa with special focus on extraction. He is currently working on mining and the chemicalisation of landscapes in colonial Zimbabwe.

Catarina Antunes Gomes holds a degree in Anthropology with a specialisation in Social and Cultural Anthropology as well as a master's degree in Sociology and a PhD in Sociology with a specialisation in Sociology of the State, the Law and Administration. She conducted her post-doctoral research in the intersection between sociology and post-colonial studies at Centro de Estudos Sociais, Coimbra University, Portugal (*The Broken Mirrors of Narcissus: dialogues between history and memory in postcolonial Angola* available at www.reseau-terra.eu/article1322.html). Some of her recent publications include "Liberdade e Comunalidade: uma leitura crítica dos estudos pós-coloniais" (*In* Para além do pós(-)colonial. São Paulo: Alameda); "On freedom, being and transcendence: the quest for relevance in higher education" (*In* Kronos. Special Issue n.º 43. University of Western Cape Town, South Africa) and *Public Humanities: Thinking Freedom in the African University* (CODESRIA, 2021 – forthcoming book).

Endalkachew Hailu Guluma is currently Assistant Professor of Literature in the Department of English Language and Literature, Arba Minch University, and an AfOx Torch virtual fellow at The Oxford Research Centre in the Humanities. Besides teaching literature courses at the BA and MA levels and advising students with their theses, he is presently working on three research projects: A Study of the Traditional Environmental Management and Conservation System of the Aari People of Southern Ethiopia, Gender in Gamo Language and Orature: A Critical Discourse Analysis, and The Institution of Kare/Ogade among the Gamo of Southern Ethiopia. Endalkachew is currently working on a community-based environmental conservation project proposal to be implemented in Aariland.

D. A. Masolo is Professor of Philosophy at the University of Louisville, USA, and has published monographs, refereed journal essays, and book chapters in his discipline. In his service to the discipline, Masolo is Co-Editor of the World Philosophies Series and a member of the Editorial Board for three other Indiana University Press book series. He also serves on the editorial boards for approximately a dozen academic journals. He served as Program Committee Co-Chair for the 2015 African Studies Association Annual Meeting. Besides working on a couple of monographs, the highlight of Masolo's current focus is editing and contributing to an anthology of essays on the Ghanaian philosopher, Kwasi Wiredu.

Tafadzwa Mushonga is Research Fellow at the Centre for the Advancement of Scholarship, University of Pretoria. She is currently working on two projects. Her first project is on extractivism and the environment which seeks to understand broader resource exploitation, environmental, and socio-economic issues. Her second project is on Climate Justice and Problems of Scale examining the relationship between the effects of climate change and the intensification of injustices in the social-political spheres. Mushonga has recently coedited a book on the *State, Militarisation, and Alternatives* published by Edward Elgar.

George Odera-Outa is an Associate Professor at the Technical University of Kenya (TUK), previously at the *Institute for Climate Change and Adaptation*, University of Nairobi. He is a Law graduate of the University of London (UK) and the University of California–Hastings (USA); previously a "Tothil Prestige Scholar" at the University of the Witwatersrand (South Africa). Outa has received various prestigious awards, among them, the *St Andrews (UK) Prize for the Environment* (2001) and the *Governance in East Africa Research Awards* (1993). He is a co-investigator in the UKRI-funded, "*Transforming Social Inequalities Through Inclusive Climate Action*" (TSITICA) and TUKs institutional Coordinator for "*Capacity Building for Socially Just Energy Transitions in Eastern Africa* project" (NOHRED II). Professor Outa is also an affiliate of the *Stockholm Environment Institute* (SEI), with whom he has recently collaborated, together with the Government of Kenya to produce the "*Science, Research and Innovation for Harnessing the Blue Economy*" (2021).

James Ogude is Director of the Centre for the Advancement of Scholarship, University of Pretoria. He is Professor of African Literature and cultures, with a special focus on memory and post-colonial literatures, popular culture, and more recently, Ubuntu and African ecologies. He recently concluded a five-year project on the Southern African philosophical concept of Ubuntu funded by the Templeton World Charity Foundation, which produced four edited volumes and an Ubuntu-themed play. He is currently leading a Mellon-funded supra-national project on African Urbanities. He is also the Director of the African Observatory for Environmental Humanities located at the Centre for the Advancement of Scholarship at the University of Pretoria. He is the author of *Ngugi's Novels and African History: Narrating the Nation*. He has edited eight books and one anthology of African stories. His most recent edited volume is *Ubuntu and the Reconstitution of Community* (Indiana UP, 2019).

Emil Roduner is a former Chair of Physical Chemistry at the University of Stuttgart and since his retirement a part-time professor at the University of Pretoria in South Africa. He wrote an advanced textbook on *Nanoscopic Materials: Size-Dependent Phenomena* (2006/2014), and together with three coauthors one on *Optical Spectroscopy – Fundamentals and Advanced Applications* (2018). His broad research interests include studies on structure, size effects, and magnetism of platinum nano-clusters and dynamics of molecules in the pores of zeolites, mechanisms of elementary steps in catalysis, and proton conductivity of fuel cell polymer membranes. At the University of Pretoria, he helped building up activities on the electrochemical conversion of CO_2 to liquid fuels using solar energy, which is of particular interest to South Africa. Furthermore, his retirement status allows him to work on fundamental questions such as the origin of the irreversibility of thermodynamic processes or the origin and nature of life.

Egmont Rohwer is the former Head of the Department of Chemistry at the University of Pretoria. He specialises in Mass Spectrometry and Chromatography to establish the chemical composition of complex mixtures. This is done to address research problems in fields ranging from engineering, biology, geology, and archaeology to forensic, environmental, medical, and food sciences. He has recently also initiated a long-term project aiming to convert renewable energy and CO_2 to liquid fuels. He finds the greatest challenge in designing new instrumentation and developing customised analytical methods to, i.e., research pheromone communication in insects, environmental pollution, wine aroma, lubricants in synthetic diesel fuel, and diagnosis of disease. He served on the South African team that negotiated the detailed terms of the international Stockholm Convention on Persistent Organic Pollutants (POPs) and is the recipient of the Georg Forster award 2018 of the Alexander von Humboldt Foundation, recognising research relevant to developing countries.

Abbreviations

AAC	Anglo American Corporation
BEAs	Bilateral Environmental Agreements
CBD	Convention on Biological Diversity
CCBA	Climate Community and Biodiversity Alliance
CITES	Convention on International Trade on Endangered Species
CSC	Cold Storage Commission
CSR	Corporate Social Responsibility
NAZ	National Archives of Zimbabwe
ECZ	Environmental Council of Zambia
EMA	Environmental Management Act
EMA	Environmental Management Agency
EPPCA	Environmental Protection and Pollution Control Act
FC	Forestry Commission
FPU	Forest Protection Unit
FQM	First Quantum Minerals
GHGs	Green House Gases
GRZ	Government of the Republic of Zambia
HSBCP	Hwange Sanyati Biological Corridor Project
ICJ	International Court of Justice
IEC	Information, Education and Communication
IEL	International Environment Law
ITTA	International Tropical Timber Agreement
ITTO	International Tropical Timber Organization
KAZA	Kavango Zambezi
MEAs	Multi-Lateral Environment Agreements
MMD	Movement for Multiparty Democracy
NAZ	National Archives of Zambia
NCCM	Nchanga Consolidated Copper Mines
NCS	National Conservation Strategy
NGOs	Non-governmental organisations
NLBA	Non-Legally Binding Agreement
NLHA	Native Land Husbandry Act
PDC	Painted Dog Conservation

PPF	Peace Parks Foundation
RCM	Roan Consolidated Mines
RST	Rhodesian Selection Trust
TFCA	Transfrontier Conservation Area
TFA	Things Fall Apart
UN	United Nations
UNFCCC	United Nations Framework Convention on Climate Change
UNIP	United National Independence Party
WB	World Bank
ZimParks	National Parks and Wildlife Management Authority
ZCCM	Zambia Consolidated Copper Mines

1 Introduction

The Intractable Problem: Africa and the Pitfalls of Resource Exploitation in a Globalising World

Tafadzwa Mushonga and James Ogude

Introduction

On 23 October 2020, South Africa woke up to the news that a 65-year-old environmental activist was shot dead at her home in KwaZulu Natal. Fikile Ntshangase was shot for "courageously standing up against Tendele Coal Mining's expansion and speaking the TRUTH," tweeted her attorney, Kirsten Youens (2020). It emerged that Fikile, the vice-chair of the subcommittee of the Mfolozi Community Environmental Justice organisation, had "refused to sign a corrupt deal with Tendele mine" (Singh 2020). She had been one of the many that protested the extension of an opencast coal mine owned by Tendele Coal in Somkhele and close to Hluhluwe–Imfolozi nature reserve on the KwaZulu-Natal North Coast (News24 2020). But according to the Centre for Environmental Rights, the killing of Fikile was not an isolated case. In April 2020, a woman activist had been shot while another was coerced to relocate after being threatened (Centre for Environmental Rights 2020). Even Fikile's death had not been the first attempt on her life. Earlier in the year, her home had been attacked, the Mfolozi Community Environmental Justice Organisation reported to South Africa's News24 (Singh 2020). Before Fikile was killed, groundWork, Human Rights Watch, and the Centre for Environmental rights had released a report detailing the targeting of community rights defenders in various South African provinces.[1] And, just as Fikile's case is not entirely new, South Africa is part of a global plight of environmental defenders under attack. In July 2020, Global Witness reported that about 212 environmental defenders, mostly from Africa, Asia, and South America, were killed in 2019 alone (Global Witness 2020).

Tendele Coal Mining has been operating in Somkhele since 2007. It started with a single plant and has been expanding its operations. According to Steyn (2020), Tendele Coal Mining plans to expand its operation because deposits in current active mines are likely to be depleted by 2022. Such expansion, the coal mining company defends, will create 1,600 jobs benefiting 87% of the community with more than 20,000 community members benefiting from the mining operations. The expansion will further enable the mine to operate for the next 10 years and sustain a supply of coal to South Africa's ferrochrome industry. But for the Mfolozi Community, Environmental Justice Organisation's plans to expand into their community have negative implications on their environment, health, and livelihoods (Rall 2020). Although rather polarised, the community is also opposed to being moved from their ancestral land (Steyn 2020).

The case of Tendele Coal Mining and controversies over its operations resulting in the death of a community member is significant in that it represents the realities of Africa's environmentalism. Such realities are often characterised by "disagreements over meanings, values, and/or functions of land or resources lead[ing] to contrasting perspectives of a particular place, environment or landscape" (Hiner 2016: 51). The Tendele Coal Mining case further particularly speaks to an intractable problem of resource exploitation and dispossession that the collection in this volume endeavours to thread together. Our focus on resource exploitation is deliberate for two reasons. First, land in Africa has historically been state-governed for the exploitation of resources (Kwashirai 2009; Larteza and Sharp 2017). The continent continues to witness colonially inherited patterns of resource appropriation being "reinforced by post-colonial patterns of uneven and combined development and capital accumulation" (Greco 2020). Iheka (2018: 10), has made a compelling argument that a deliberate representation of Africans in colonial discourse, "as wild, to be tamed and controlled," was an attempt to naturalise the Africans in order "to justify the colonial enterprise – that is, to provide rational for the exploitation of the human and nonhuman resources in the territories, albeit in the name of civilization and progress." In the context of colonial modernity and the current global modernity, the ruthless despoliation of the African environment has rested firmly on this logic of naturalising Africa. Significantly, multi-national corporations working on the continent have continued this colonial disregard and contempt for the environment in the communities where they operate. Again, Iheka writes that "[b]y disregarding 'best practices' adopted in their home countries and in other Western countries, the companies continue to treat these African environments as devoid of people or constituted by disposable people." (13)

Secondly, resources in Africa are not limited to earth-based natural resources such as land, water, minerals, timber, and wildlife. In this volume, we consider that African people, their indigenous languages, culture, and knowledge systems are in themselves resources predisposed to all forms of misreadings and deletion. When this happens, an ontological chasm is created between culture and nature, humans and non-humans – an artificial nature/culture dichotomy is created. Therefore, in indigenous communities where a multi-species co-existence was encouraged, a deliberate deletion or wilful repression took place under colonialism. When African communities begin to engage in resistance under environmental justice, it is often without any ontological anchor rooted in local ways of knowing. Instead, environmental justice struggles tend to borrow their language for framing issues from the perspective of the so-called green revolution in Europe and North America.

Over the years, due to its vast resources, Africa has been subjected to what Harvey (2005) describes as accumulation by dispossession. Accumulation by dispossession is a dominant feature of most African economies. It is fraught with environmental, cultural, racial, and several other injustices related to the general wellbeing of people (Larteza and Sharp 2017; Mitman 2017). Such injustices are often hidden in what Fairhead and colleagues call "green credentials" – justifications to appropriate land driven by environmental agendas (see Fairhead et al. 2012). Accumulation by dispossession also often comes hidden in development rhetoric. For instance, Ayelazuno (2014) demonstrates how resource exploitation projects are viewed as development projects with promises for economic benefits and jobs at the local level. Such justifications have strategically earned powerful public and private actors the legitimacy to exploit an already disadvantaged population who has no choice but to depend on any job that a dispossessing industry will offer (Hallowes and Munnick 2016; McKay 2020). So, while natural resources in Africa should benefit the African people, they have become the very means to their marginalisation or what Rob Nixon (2007) calls slow violence against the poor. Populations across Africa are predisposed to landlessness, joblessness, homelessness, food insecurity, social disintegration, and many other forms of human rights violations (Robinson 2003).

Set within the broader understanding of the resource relationship between the industrialised global north and the developing global south, our general argument is that the dispossessing nature of resource exploitation, despite justifications, is at the core of what defines the perverse nature of Africa's environmental struggles. These struggles revolve around environmental (in)justice, violent capitalocenes, repression of indigenous knowledge, and climate change. Because we are living in the Anthropocene – the geological epoch in which human beings are the main cause of the environmental change (Crutzen and Stoermer 2021), we deliberately focus on human societies, their human ways of being, and the resulting intricate relations with their environments. In that regard, we ground our analysis of Africa's environmental struggles within the broader Environmental Humanities – understanding environmental issues from within the humanities and social sciences. We, therefore, in part, adopt what we call an Environmental Humanities of Extraction (EHfE), a framework we believe helps us better grasp the intricate web of African extractive geographies, and embed such complexity in extant and diverse ways of understanding contemporary environmentalism. The Environmental Humanities of Extraction approach that we emphasise in this book, and which is an important aspect of broad environmental exploitation, follows the realisation that any endeavour to understand resource extraction and its consequences on the continent must be done within an understanding of nature and human ecology; science and the environment; land and de-coloniality; economics and marginality, and broader issues around land, politics, development, and infrastructure.

Succeeding chapters offer diverse perspectives from a range of fields in the humanities and social sciences, in essence, departing from the single focus on disciplines to a common effort in disentangling the complexities of today's environmental crises (Sörlin 2012). The chapters further use examples from different geographical locations on the continent. In themselves, these chapters are embedded within the basic principles of an environmental humanities framework. For now, we emphasise a broad overview of how each of Africa's environmental challenges can be traced back to resource exploitation. In doing so, we rely on the Humanities in bringing "questions of meanings, value, ethics, justice, and politics of knowledge production" into the environmental domain (Rose et al. 2012: 2), more importantly in bringing humanistic modes of critiquing dominant narratives about humans and nature (Neimanis et al. 2015). While we discuss these themes distinct to draw attention to the range of humanities issues they open up, we acknowledge that, in reality, these problems manifest in complex interrelationships.

Environmental Justice in Africa

The name "Mfolozi Community Environmental Justice Organisation" is suggestive of a permeating environmental crisis in which justice has become the centre of the community's struggle. Dating to as many as five decades back, movements like the Mfolozi community have popularised the "environmental justice" concept in contempt of wide-ranging environmental concerns, particularly the equal distribution of environmental costs and benefits from various resource exploitation projects (Jenkins 2018). Apart from the distributional issues linked to exploitation projects, these movements also often call for fair treatment and equal opportunities in decision-making regarding the implementation of such projects regardless of race, gender, economic class, or ethnicity (Holifield 2001; Sovacool and Dworkin 2015; Jenkins 2018). Just compensation has also become part of the call for environmental justice (Čapek 1993). The moral dimensions of environmental justice have, therefore, over the years, encompassed processes of distributive, procedural, and restorative justice (Čapek 1993; Holifield 2001; Schlosberg 2013; Sovacool and Dworkin 2015), which include broader merits of social justice as part of an environmental humanities enquiry. And within the environmental humanities, calls for environmental justice also demand for multi-species justice as humans and non-human species often coexists in their environments (Robin 2018).

Over the years, environmental justice has been adapted to several aspects of resource exploitation and the environment. Energy justice, for instance, calls for equal distribution of costs and benefits from energy systems and fairness in the decision-making processes regarding the planning and implementation of energy projects (Sovacool and Dworkin 2015). Similarly, climate justice emerges largely out of the climate crisis and concerns around sharing the costs and benefits of climate change mitigation, adaptation, and reducing carbon emissions. Critical questions of the climate justice debate include, who benefits from the emissions of carbon dioxide? Who should bear the burden of mitigation? Who should cover the cost of adaptation amongst the world's rural poor? (see Jenkins 2018). These questions arise as environmental humanists critique the history of colonialism and its corollaries on indigenous and minority people in continents like Africa (Adamson 2016). The same principles and questions guide debates around conservation, water, and land justice.

At the core of environmental justice struggles in Africa is the dispossessing nature of water, oil, minerals, timber, wildlife, and fisheries exploitation, which are often marked by different forms of environmental degradation, land dispossession, corruption, food insecurity, and weak consultative processes with resident communities (Cock 2015). Hallowes and Munnick (2016) of groundWork in South Africa, for example, show that mega extraction projects are, by their nature, breeding grounds for environmental injustice. This injustice, they argue, is imposed on people in three ways. The first is by degrading people's environments, including the degradation of their very own bodies as they provide cheap labour. Secondly, injustice is imposed on people by territorialisation of common or public goods. The case of violent exclusion from resources in Zimbabwe's Sikumi Forest Reserve discussed in Chapter 4 of this volume speaks to this form of injustice. It draws attention to the colonial territorialisation of common goods such as forests, accentuating the political processes of how injustices of resource alienation are sustained in a post-colonial state. Thirdly, Hallowes and Munnick demonstrate that people are subjected to injustice when they are excluded from political and economic decision-making processes often resulting in serious disadvantage and a general state of dispossession.

In addition to these views, we do also recognise that today's struggles for environmental justice are set within shifting historical meanings of the environment. The advent of colonisation changed the meanings of nature and the environment in ways that supported an insatiable need for colonial capital. The environment was seen as a source of wealth and development. As a result, the main agenda of colonial conquest became the exploitation of raw materials for economic expansion (Larteza and Sharp 2017), and state making (McQuade 2019). For African indigenes, the environment remained a place where they lived with nature for mutual benefits (McGregor et al. 2020). The new ways of viewing the environment promoted a disenfranchising colonial legacy whose agenda also became to erase African worldviews about the environment. African literature, such as the works of Ngugi wa Thiong'o discussed in Chapter 6, has been adept at expressing the legacy of colonialism in environmental degradation and displacement of African environmental ontologies with an imperial discourse, and alienation of people from their material and spiritual being, which is inextricably linked to nature and the totality of their planetary universe. The environment having acquired new meanings in the wake of colonialism, forests that supported diverse non-human lives, and the lives and cultures of indigenous communities were replaced by plantation economies, which in turn replaced traditional livelihoods and culture with disease, slavery, landlessness, and polarisation amongst indigenous Africans (Mitman 2017). Communities were evicted from their homes of centuries in favour of timber, mineral, and oil extraction (see Hilson 2002; Butler 2006; Kwashirai 2009; Larmer and Laterza 2017), fortress conservation (Brockington and Igoe 2006; West et al. 2006; Brockington et al. 2008) and were introduced to ways of living steeped in racial and economic injustice, and subjugating poverty.

Overall, contrasting views on the meanings of the environment formed the basis of contested ecologies (also see Nygren 2004; Green 2013), which largely influence the way discourses on environmental justice have developed. The growing number of environmental justice movements over time demonstrates the temporality of environmental and social justice crises on the continent. From decade to decade, we see a production of colonial legacies on the post-colonial moment – a continuation of a neo-colonial trajectory in insidious ways that disenfranchise citizens under the guise of development. We, therefore, think that unjust development should be conceptualised as one of the major challenges defining Africa's environmental crisis.

The Proliferation of Violent Capitalocenes

The proliferation of environmental justice movements is not an entirely surprising phenomenon for a continent where the exploitation of resources is historically marred by violence and an insatiable need for capital. We generally think of the problem of capital, development, and different forms of violence associated with resource exploitation in terms of what we call "violent capitalocenes."

The term "capitalocene" emerged as an alternative viewpoint of the Anthropocene debate, which recognises the global ecosystem as severely altered by human beings to the extent that the current geological epoch should be recognised as "one characterised by a pervasive human signature" (Buscher and Fletcher 2020: 18). Others think of the Anthropocene as the "new human condition" describing it as "the unprecedented crisis of how we as a species will cope with the consequences, not to mention responsibilities of being the major driver of planetary change" (Holm et al. 2015: 983). While still pursuing an argument on the ecological crisis, Jason Moore, a strong critique of the Anthropocene does not think we are facing an anthropogenic ecological crisis, but rather a capitalogenic one (Moore 2017, 2018). In that respect, he argues that we are living not in "the age of man" rather in the "age of the capital" – the capitalocene. As such, the planetary crisis needs to be understood in the context of the world-ecological power of capitalism, which, unlike the Anthropocene concept, is not bound by geological epochs.

Using Jason Moore's theoretical perspectives on the Capitalocene, our conceptualisation of violent capitalocenes emphasises a form of capital accumulation entrenched in violent modes of resource exploitation. It points to the convergence of resource extraction, capital accumulation, and violence. And by violence, we do not only limit ourselves to direct inflictions of harm on human and non-human resources. Like Galtung (1969), we believe the violence of resource exploitation to a large extent also manifests in systemic and symbolic ways.

We argue that there is no better place to start having a debate on the cumulative effect of capitalism other than the continent of Africa. Indeed, the manifestation of "violent capitalocenes" is most stark in Africa and continues to shape north/south geopolitical relations. Common on the continent is petro-violence – violence and conflict resulting from oil exploitation (Watts 1999; Zalik 2004; Finer et al. 2008; Obi 2014). Also widespread is conservation violence perpetrated in the guise of saving biodiversity from extinction (Duffy 2000; Neumann 2004; Brockington and Igoe 2006; Büscher and Ramutsindela 2016). This is the kind of violence that environmentally aware humanities scholars like Njabulo Ndebele (1998) intimates when he critiques the creation of game lodges and nature reserves as the anti-thesis of nature conservation, but rather a form of leisure reserved for whites, the type which, for a black middle-class tourist can only elicit what Rob Nixon (2007), calls, "a suppressed history of dispossession." We see in the work of both Njabulo Ndebele and Rob Nixon, and several scholars in the field of political ecology, geography, and history the critical importance of the environmental humanities in challenging established truths or "unsettling of dominant narratives" (Rose et al. 2012: 3; Sörlin 2012) about humans and their environments. Violence in the mining sector is also pervasive (Katsaura 2010; Hallowes and Munnick 2016), and so is social violence from infrastructural projects such as the Lesotho highlands water project (Robinson 2003; Mwangi 2007).

Cases of "extreme poverty, environmental degradation, human rights abuses, authoritarianism, civil conflicts, and wars are rife" in almost every form of resource exploitation in Africa (Duruigbo 2005: 2). The tragedy of endowment (Alao 2007), paradox of plenty (Karl 1997), the resource curse (Ross 1999), and the idea of resource scarcity (Homer-Dixon 1994), endeavour to theoretically understand why a continent abundantly endowed with resources experiences underdevelopment and violent conflict. In search of this understanding, the spectre of coloniality cannot be overlooked. This history extends into the post-colony through the very nature of state power in post-colonial Africa, which has basically assumed state structures of the colonial state before it. With little to no adjustments, the post-colonial legal framework itself still largely supports colonial legacies of state capital. African literature, as Ogude argues in Chapter 6, has been a witness to this colonial pattern of economic exploitation of local resources – a culture that continues to be reproduced in the post-colony, and as he points out, colonial plunder of resources is linked directly to ecological devastation and deforestation in Africa. Consequently, Outa in Chapter 10 argues for legal instruments that reflect on Africa's context and needs, especially in protecting goods of the commons to the benefit of all. The political tension around the Nile River is a good case in point where colonial treaties governing the use of the Nile River are being challenged to accommodate all states that have a stake in it.

Moreover, we think that the colonial model of state power, which thrived on a deliberate weakening of traditional structures of authority, continues to play itself out, thus weakening the governance structures within most African states (Mahmood 1996). Weak governance structures render the continent prone to infiltration by former colonisers who still have an agenda to benefit from the continent's rich resource base. Their influence ensures a kind of politics that leaves the control and exploitation of resources at the behest of a few powerful national and international actors. Developments in Zambia's Copperfield explored by Chansa in Chapter 5 are demonstrative of this kind of state politics, wherein former colonisers come as foreign investment, while the state uses such investment for political gains. As a result, Obi (2007) argues that resources alone do not lead to violence, rather the politicisation of resources does. In this volume, we likewise recognise the politicisation of resources as one of the principal drivers of violent capitalocenes and part of the broader nature of Africa's problems. It is by no means a surprise that de-politicisation of resources is one of the key agenda items for environmental justice movements such as the one in South Africa's Mfolozi Community.

Indigenous Knowledge

Violent resource exploitation has often resulted in the alteration, modification, or erasure of indigenous knowledge, practices, and ontologies. We understand indigenous knowledge as the local knowledge, ideas, and practices accumulated over a long period through continuous interaction, observation, and experimentation. Such knowledge, ideas, and practices are embodied in a community's ways of being and becoming; ways of thinking through their identity and acts of survival in relation to their health, food security, and the overall management of their natural resources (Warren 1991; Mundy 1993; Green 1996; Moahi 2007; Dutta et al. 2018). For this reason, indigenous knowledge is considered as social capital (Ajayi and Mafongoya 2017; Masolo 1994), and therefore a resource for local communities (see, e.g., Boillat and Berkes 2013).

We have already argued that the western and colonial idea of nature is very much linked to the environment as a commodity and resource for industrialisation, largely considered as the sole vehicles for development and job creation (Dutta et al. 2018). Indigenous worldviews of the environment, on the other hand, are embedded in the spiritual idea of "earth keeping" rather than "earth exploiting." The Aari people of Ethiopia whose environmental ontology is explored in Chapter 8, for example, think of the environment as not only consisting of animals and plants, but one that also encompasses the gods, ancestors, and humans in constant interaction with one another. The sustainability of this complex relationship is founded upon the acknowledgement of their mutual interdependence – hence their philosophy of nurturing rather than exploitation of the environment based on the principle of co-agency and basic minimum harmony with nature. The idea of earth keeping is, thus, according to Holm et al. (2015), a way of critiquing local and global practices linked to capitalism that contribute to negative changes in people's traditional ways of living, and to environmental change. It becomes, in itself, a humanities analytical approach for responding to problems of both social and environmental change in Africa.

When we first touched on the different meanings of the environment, we aimed to highlight the origins of contested ecologies and the rise of environmental justice movements. On a different note, but with the same trajectory, we also note how hegemonic development discourses founded on resource exploitation, or manipulation in the case of culture and tradition, have reconstructed indigenous ontologies, epistemologies, and philosophies (Agrawal 1995; Moahi 2007; Legesse et al. 2013; McGregor et al. 2020). Ayodeji Adegbite's Chapter 7, for example, draws our attention to how colonial and African elite reconstructions of Yoruba identities in Nigeria amounted to a violation of indigenous ontologies and epistemologies, resulting in new forms of relationships with the land. A good example is the transformation of land into the property and a commodity that could be acquired only through legal tender, which in turn led to land appropriation and the emergence of a violent plantation industry.

It is also important to highlight that over 50% of the world's natural resources are located on land already settled by indigenous people (Terminski 2013). When oil or gold is discovered on such land, or when land is found suitable for monocultural tree-crop production such as in the case of rubber and cocoa, African states will often consent to displace people in favour of development, geopolitical ties, and political power. Chansa's chapter illustrates how the establishment of mining facilities in Zambia's Copperbelt has resulted in the displacement of local communities and distorted indigenous form of subsistence. Such examples, supported by insights from Legesse et al. (2013), demonstrate that development-induced displacement systematically disintegrates communities together with their wealth of knowledge accumulated over many generations. Dutta et al. (2018) argue that any form of displacement from a community's natural habitat often leads to the destabilisation of social structures forming the very core of indigenous knowledge resources. Gradually, indigenous practices are weakened, sometimes altogether eroded, and the problem is compounded by a lack of efficient documentation of indigenous forms of knowledge as compared to modern modes of knowledge (Eyong 2007).

The dearth, even death, of indigenous knowledge systems is part of Africa's intractable problems in three ways. First, contrary to the justification that resource exploitation is a development strategy whose objective is to bring economic benefits to people, the structural and political processes involved often result in the weakening and erasure of once resilient indigenous systems (Dutta et al. 2018). Instead of development, communities are left to endure gradual erasure of tradition and culture, rights to land, livelihoods insecurity, and impoverishment. Land is the most contentious issue in Africa (see, e.g., Ogude 1999; Moyo 2003; Takeuchi 2022). As we have highlighted earlier, landlessness is a common consequence of resource exploitation. We know from Cernea (1999) that land is the foundation upon which production systems are constructed. We also know that land, as place, is part and parcel of the cumulative indigenous knowledge base providing for a society's resilience to environmental change (Ford et al. 2020). The ontological value of land among African people, as demonstrated in Chapter 6, continues to be a major pre-occupation of African writers of fiction like Ngugi wa Thiong'o. Land appropriation for development projects, therefore, destroys the very basis of life, the outcome of which is livelihood insecurity and impoverishment as we argue in some of the chapters. Land-based conflicts persist with land appropriation, not only weakening the wellbeing of people but also costing their lives like in the case of Tendele Mine and the death of Fikile Ntshangase.

The second relates to the way erosion of indigenous knowledge systems has an impact on sustainable development. The "Our Common Future" report defines sustainable development as "development that meets the needs of the present without compromising the ability of the future generation to meet their own needs" (Brundtland 1987:8). Although the report has been in some spheres contested as a regressive document (Trainer 1990), the fact that it connects development to environmental health provides useful insights for interrogating the crisis of resource exploitation that we grapple with in this volume. Institutional arrangements and social structures that produce indigenous knowledge have not only supported cooperation and exchange of survival skills within and between communities; they have ensured the sustainability of those communities and laid the foundation for the survival of their futures. But resource exploitation and associated displacements hypothetically tear into the local social fabric disabling the sustainability of existing survival tools. Like Eyong (2007), we think that development based on the exploitation of resources is a threat to the sustainable development of particularly vulnerable communities in Africa in the way it distorts the sustainability of indigenous knowledge and practices.

The third pertains to the ecological crisis. Ecological crisis is not only a problem because of the rampant resource plunder, but also because the plunder of resources is eroding indigenous knowledge, which has, for over many decades, provided a distinct diagnosis for ecological crisis (McGregor et al. 2020). As a way of life, many communities in Africa are accustomed to monitoring, responding, and managing ecosystem processes and functions, which have given both human and non-human systems resilience and adaptive strength to overcome environmental shocks (Berkes et al. 2000; Ajayi and Mafongoya 2017). There is increasing recognition that traditional knowledge contributes to conservation and sustainable resource use (Gadgil et al. 1993; Hens 2006; Adom and Kquofi 2016; Mavhura and Mushure 2019). There is, furthermore, an increasing awareness that the ecological crisis of our time needs the coming together of modern and indigenous knowledge (DeWalt 1994; Mercer et al. 2010; Mistry and Berardi 2016). Again, the example of the Aari people's traditional and religious beliefs, and their linkages to the environment in southern Ethiopia explored in Chapter 8 acknowledges threats to indigenous knowledge and argues for the integration of indigenous with modern knowledge for sustainable and community-led environmental management. We, however, think that in geographies like Africa, where the need for capital is domineering, and the use of violence in the exploitation of resources increasing, it is becoming increasingly difficult for communities to accumulate and use their knowledge to protect themselves and their environments. The overall point we make is that resource exploitation may have positive outcomes for national growth and geopolitical economic relationships, but certainly negative impacts on indigenous knowledge systems, and the capacity of local people to respond and adapt to environmental shocks. We think that these negative impacts are significantly one of the major pitfalls of resource exploitation on the African people.

The Problem of Climate Change in Africa

According to the World Meteorological Organisation (WMO) report (World Meteorological Organisation 2020), 2019 was the third-warmest year on record following 2010 and 2016, with temperatures exceeding 2°C above the 1981–2010 average being recorded in South Africa, Namibia, and parts of Angola. Many areas from south to north of the continent were 1°C above the normal temperature. A continued rainfall deficit was recorded in Southern Africa. East Africa experienced erratic rainfall throughout the greater part of 2019, while flooding affected various parts of the Sahel, with Sudan and the Central African Republic being severely affected. The WMO predicts that large parts of Africa will exceed 2°C of warming to the 20th century's mean annual temperature by the last two decades of this century. These predictions are consistent with those reported by Africa Partnership Forum (2007), that most of the Sahel and southern Africa will experience warming in the range of 3–6°C by 2100, and congruent to a decline in precipitation by more than 20%. It is believed that by 2080, areas under arid land in Africa will likely increase by 5–8% (Williams 2012). Climate change is, therefore, considered to be an increasing threat to Africa, posing a great concern for the water resources, agriculture, health, ecosystems, and biodiversity sectors (Africa Partnership Forum 2007; Williams 2012; World Meteorological Organisation 2020).

Although temperature and rainfall statistics, and predictions for the future are certainly a cause for concern, we argue that the climate change problem in Africa lies not in these empirical figures. Much of the problem lies in the uneven geographical distribution of climate change effects better analysed and defined through a critical humanities enquiry (Holm et al. 2015). Compared to other regions in the world, Africa is one of the continents severely affected by the projected warming and sporadic rainfall patterns. What makes Africa particularly exposed to the effects of climate change is the vulnerability of its population (Thompson et al. 2010). This vulnerability is primarily due to the continent's dependency on natural resources for survival and is severely felt across disadvantaged categories of society. Rainfed agriculture, for example, is a source of livelihood for more than half of the population and is directly dependent on favourable climate conditions (Africa Partnership Forum 2007; Collier et al. 2008; Lipper et al. 2014). So, any environmental stress related to climate change has adverse outcomes on livelihoods, human, animal, and crop health, and agro-based economic activities (Thompson et al. 2010; Lipper et al. 2014; World Meteorological Organisation 2020).

Because Africa depends on natural resources, vulnerability to climate change is further increased by widespread environmental degradation. The dependency created by capitalocenes on high carbon-emitting sources of energy and the destruction of the environment reduce the environment's capacity to meet social and ecological needs (UNEP undated). Loss of ecosystem services increases exposure and susceptibility to the effects of climate change, which, in turn, often result in further environmental degradation. Pre-existing socio-economic and socio-political constraints, steeped in colonial injustices and failure by post-colonial governments to control the expansion of violent capitalocenes, add another layer of difficulty in adapting to or mitigating the adverse effects of climate change (Hallowes and Munnick 2016; Larteza and Sharp 2017; Hallowes and Munnick 2018).

Hence, it is the poorest on the continent that will bear the brunt of climate change (World Meteorological Organisation 2020). The majority of people constantly referred to as the poorest on the continent, are those living in rural areas[2] where oil, minerals, and other valuable resources are being discovered and exploited without clear regulatory frameworks for environmental protection (Terminski 2013). Unregulated expansion of economic projects, tied to resource exploitation, into rural lands increases not only the risk of losing land and the wealth of knowledge attached to the place, as we have argued before. It increases vulnerability and the very risk of failing to cope with climatic uncertainties. More compounding is that contemporary ideas on development continue to heavily depend on the exploitation of resources (Wood and Roelich 2019). For example, the coal industry continues to flourish in most parts of the world, including Africa (Hallowes and

Munnick 2018; Brown and Spiegel 2019). Even more aggravating is that climate change itself induces displacement, which, in turn, further destabilises communities and indigenous knowledge systems. In some instances, displacement as a result of climate change is believed to undermine peace and stability, also affecting people's ability to cope (Africa Partnership Forum 2007). For example, South Africa has been grappling with migration-fuelled violent xenophobia attacks, which some link to climate change (Chikulo 2014; Klaaren 2021). Approximately, 60% of internal displacement in the East and Horn of Africa during 2019 was due to climate disasters resulting in pastoralists being forced out of pastoralism into displacement camps (World Meteorological Organisation 2020).

The problem of climate change in Africa should also be conceptualised as a problem of scale and a case of geopolitical environmental injustice. The distributional impact of climate change is such that rich countries responsible for the highest amount of carbon emissions and global warming are less affected by the impacts of climate change. Poor countries in the global south, on the other hand, are bearing the brunt. Within Africa, the Covid-19 global pandemic has bluntly reminded the continent of systemic racism, economic deprivation, and general social stratification (Chirisa et al. 2020; Obeng-Odoom 2020; Visagie and Turok 2020), which are, to a large extent, linked to resource exploitation and the uneven distribution of climate change effects across societal scales of geography, gender, economic class, race, ethnicity, and age (Islam and Winkel 2017; Brown and Spiegel 2019). Therefore, advocates of climate justice movements, and scholars alike, are increasingly calling for distributional justice between developed and underdeveloped countries, low- and high-income households (Bou-Habib 2019; Puaschunder 2020; Nyiwul 2021).

Another pervasive problem in Africa is, however, that governments and the corporate world are often not in good relations with people who call out for environmental and climate justice. Lack of trust, cooperation, and accountability often obstruct the path to peaceful and meaningful engagements leading to further injustices (Whyte 2020). Another layer to these challenges is how authoritarian regimes often perceive the struggle for environmental justice, or any form of justice for that matter, as political dissent and an act of treason. Evidently lacking is the capacity for leadership and moral authority around which most societies can rally as they grope for a grammar to frame vexed climate issues. Masolo's chapter speaks directly to this challenge, pointing out lessons of moral leadership during periods of climatic crisis that can be drawn from rainmaking traditions in Kenya. It is against these perspectives that we think the climate change problem in Africa goes way beyond the scientifically generated facts of carbon emissions and global warming. For us, and from a broader environmental humanities perspective, the problem is entrenched in the pervasiveness of colonial legacy, capitalism, and industrialisation dependent on resource exploitation, and the associated scalar issues looming large over Africa.

Conclusion

The general argument developed in the range of chapters that follow is that resource exploitation and its embeddedness in dispossessing colonial ideas of capital are linked to what defines the nature of Africa's problems. The chapters are predominantly African voices from the humanities expanding on the four themes introduced in this introduction, but also bringing out other issues related to legal frameworks, leadership, as well as connecting climate science with broader social dynamics related to energy use and transition.

Catarina Gomes's chapter pitches the tone of the volume by asking difficult but pertinent ontological questions on human exceptionalism. It offers a critique of Anthropocentrism in relation to human/nature relations. Set within Angola, this philosophically rich chapter is a valuable contribution to this volume in the way it triggers us to question and understand the ethical and moral position of human-centred resource exploitation. An emerging key argument is a recognition that human development is primarily about being more, rather than having more. It is about caring beyond possession or self-referentiality that still dominates our interaction with the planet. It is about the realisation of our entanglement, as human, with the non-human; the collapsing of a hierarchical arrangement of life forms in ways that privilege the human. In many ways, Gomes' chapter signals the tension often generated between environmental justice advocacy, which centres the human, while ignoring connections between species and ultimately sustainable lifeworlds without which environmental justice itself is impossible.

Elijah Doro provides the historical context of dispossession and offers a typical example of the colonial marginalisation of African people. By tracing the history of goat farming in Zimbabwe, the chapter makes deep inferences about how environmental racism, disenfranchisement, and marginalisation of people unfolded at the behest of colonial agendas. It helpfully provides a historical context of understanding the present and the future of Africa's environment and society.

From historical perspectives, Tafadzwa Mushonga focuses on contemporary resource exploitation and dispossession in conservation areas. Using the case of protected forests in Zimbabwe, she demonstrates how conservation in Africa is beset by politics of exclusion from resource utilisation, and how such exclusion is characterised by violence. In addition to being an account of the spectre of coloniality, and spatiality of violence in conservation spaces, the chapter is an illustration of the daily socio-environmental struggles of resource-dependent people, and the hegemonic role of the state in perpetuating such struggles.

Chibamba Jennifer Chansa's chapter provides a chronology of the environmental impacts of mining and its management in Zambia's extractive industry. Using the case of Zambia's Copperbelt, the chapter illustrates how environmental impacts of mining are interconnected and aligned with Zambia's political economy, which post-independence, has become heavily reliant on foreign investment with little attention given to addressing environmental degradation and its cascading effects on communities. As a result, one of the most significant mining regions of Africa is beset by poor environmental management. What is significant about this chapter is that it epitomises the politics of state-controlled mining and the diverse environmental challenges faced by mineral-rich countries across Africa.

James Ogude brings a literary perspective arguing that the issue of environmental degradation and social justice such as those illuminated in Mushonga and Chansa's chapters are critical to the extent that they have always been a major pre-occupation of African literature. Using the example of some of Ngugi wa Thiong'o's texts, the chapter argues that we cannot separate the history of the empire from ecocritical thought because it de-historicises nature by condoning the unsustainable exploitation of land and seeing African land, not as a resource to be protected, but a surplus for exploitation. The chapter also demonstrates how Ngugi offers a critique of modernity, especially how the advent of colonial capitalism leads to degraded earth and total annihilation of indigenous forms of enterprise that rested on sustainability and oneness with nature, thus disrupting the human–nature dichotomy. Ogude underscores the fact that many African societies, in spite of their complexities and differences, are drawn to an ethics of the earth, similar to what Ngotho in Ngugi's *Weep Not, Child*, evinces. As the chapter argues, Ngugi's novels bring out a form of violence towards the earth occasioned by colonial exploitation that leaves Illmorong, the setting of *Petals of Blood*, drought-stricken and unable to support its people. Ogude unpacks how Ngugi links ecological and environmental degradation to the emergence of nascent capitalism in Kenya symbolised in the steam engine that comes with rail, which must be fed wood from the forest.

Land is such a valuable resource for Africa. The idea of an Africa with "surplus land" is also taken on by Ayodeji Wakil Adegbite, who opens a discussion on human–nature relationships before and after colonialism using the setting of Yoruba identities in Nigeria. The chapter argues that as opposed to the colonial view of seeing land as surplus free for exploitation, indigenous communities have a particular relationship with land and all resources – a relationship embedded in indigenous cosmology, and one that compels people to care for rather than exploit nature. The chapter then shows how such relationships have been distorted and rearranged to suit exploitative needs at a cost only endured by Nigeria's vulnerable categories of society.

And indeed, the attitude towards land as a surplus commodity in the wake of colonialism also signals the beginnings of a philosophical disconnect from indigenous sociality to the environment that guided most traditional societies in Africa. Hailu Guluma and Zewde Jagre take us through yet another cosmological experience of the Aari of Ethiopia demonstrating the significance of indigenous ecological norms as a resource for shaping community-led conservation. The chapter importantly demonstrates the extent to which indigenous communities' ontology and rituals are a key resource for convivial principles of care and co-responsibility, but which have also been on the brink of being taken over by alternative philosophies that encourage the ethos of possession and domination over natural resources. Referencing local knowledge formations, both living and repressed, is what Catarina Gomez refers to as petitioning the future, an act of gazing into the past because "in the past, there was the future."

Drawing on the local mythology of rainmaking among a peasant community in the Western region of Kenya, Masolo's chapter demonstrates that the resource exploitation crisis in Africa is not just about domineering discourses. It is also about leadership – moral and ethical leadership – whose legitimacy must be rooted in the community's experiences and grammar of framing things. For Masolo, good leadership in Africa is one that remains focused on Africa's context and needs rather than one aimed at satisfying some geopolitical green agendas.

Yet, because of our interconnectedness as continents and countries, it is crucial to remember that no unilateral solutions are sustainable unless the world community is united in building legal frameworks within which comprehensive and sustainable regimes for protecting the earth are put in place. George Outa's chapter grapples precisely with regulatory protocols and green agendas that have been left to the dictates of a few influential countries in the Global North and China. The chapter emphasises taking leadership concerning legal framework reform. Outa argues that creative legal solutions embedded in culture and indigenous knowledge prohibit exploitative use of mother nature, and that protection of national sovereignty among the developing nations is critical to protecting resources. However, the principle must be balanced with the protection of common goods.

Emil Roduner's chapter recognises that much of the environmental crisis in Africa is irrevocably linked to the continent's carbon footprint. The chapter calls for alternative uses of energy, suggesting practical methods of how efforts from Science and Humanities can work together to mitigate the effects of climate change.

Notes

1 https://cer.org.za/reports/we-know-our-lives-are-in-danger.
2 Rural poverty Report 11 (International Fund for Agricultural Development).

References

Adamson, J. 2016. Introduction: Integrating knowledge, forging new constellations of practice in the environmental humanities. In *Humanities for the Environment*, ed. J. Adamson and M. Davis, 17–33. London: Routledge.

Adom, D. and S. Kquofi. 2016. The high impacts of Asante indigenous knowledge in biodiversity conservation issues in Ghana: The case of the Abono and Essumeja townships in Ashanti region. *British Journal of Environmental Sciences* 4:63–78.

Africa Partnership Forum (APF). 2007. *Climate Change and Africa*. 8th Meeting of the Africa Partnership Forum, Berlin, Germany, May 22–23.
Agrawal, A. 1995. Dismantling the divide between indigenous and scientific knowledge. *Development and Change* 26:413–439.
Ajayi, O. and P. Mafongoya. 2017. *Indigenous Knowledge Systems and Climate Change Management in Africa*. Wageningen: CTA.
Alao, A. 2007. *Natural Resources and Conflict in Africa: The Tragedy of Endowment*. Rochester, NY: University of Rochester Press.
Ayelazuno, J. A. 2014. The 'new extractivism' in Ghana: A critical review of its development prospects. *The Extractive Industries and Society* 1:292–302.
Berkes, F., J. Colding and C. Folke. 2000. Rediscovery of traditional ecological knowledge as adaptive management. *Ecological Applications* 10:1251–1262.
Boillat, S. and F. Berkes. 2013. Perception and interpretation of climate change among Quechua farmers of Bolivia: Indigenous knowledge as a resource for adaptive capacity. *Ecology and Society* 18:21–33.
Bou-Habib, P. 2019. Climate justice and historical responsibility. *The Journal of Politics* 81:1298–1310.
Brockington, D., R. Duffy and J. Igoe. 2008. *Nature Unbound: Conservation, Capitalism and the Future of Protected Areas*. London: Earthscan.
Brockington, D. and J. Igoe. 2006. Eviction for conservation: A global overview. *Conservation and Society* 4:424–470.
Brown, B. and S. J. Spiegel. 2019. Coal, climate justice, and the cultural politics of energy transition. *Global Environmental Politics* 19:149–168.
Brundtland, G. 1987. Our common future: Report of the world commission on environment and development. United Nations Commission. In *United Nations*. Geneva: United Nations.
Buscher, B. and R. Fletcher. 2020. *The Conservation Revolution: Radical Ideas for Saving Nature Beyond the Anthropocene*. London: Verso.
Büscher, B. and M. Ramutsindela. 2016. Green violence: Rhino poaching and the war to save Southern Africa's peace parks. *African Affairs* 115:1–22.
Butler, D. R. 2006. Environmental and social change in southwestern Sierra Leone: Timber extraction (1832–1898) and rutile mining (1967–2005). PhD diss., Texas State University.
Čapek, S. M. 1993. The "environmental justice" frame: A conceptual discussion and an application. *Social Problems* 40:5–24.
Centre for Environmental Rights. 2020. October 24. *https://twitter.com/Centre EnvRights* (accessed December 4, 2020).
Cernea, M. M. 1999. Why economic analysis is essential to resettlement: A sociologist's view. *Economic and Political Weekly* 34:2149–2158.
Chikulo, B. C. 2014. An analysis of climate change, poverty and human security in South Africa. *Journal of Human Ecology* 47:295–303.
Chirisa, I., T. Mutambisi, M. Chivenge, E. Mabaso, A. R. Matamanda and R. Ncube. 2020. The urban penalty of COVID-19 lockdowns across the globe: Manifestations and lessons for Anglophone sub-Saharan Africa. *GeoJournal* 2020:1–14.
Cock, J. 2015. How the environmental justice movement is gathering momentum in South Africa. *https://theconversation.com/how-the-environmental-justice-movement-is-gathering-momentum-in-south-africa-49819* (accessed January 16, 2012).
Collier, P., G. Conway and T. Venables. 2008. Climate change and Africa. *Oxford Review of Economic Policy* 24:337–353.

Crutzen, P. J. and E. F. Stoermer. 2021. The 'Anthropocene' (2000). In *Paul J. Crutzen and the Anthropocene: A New Epoch in Earth's History*, ed. P. J. Crutzen, 19–21. Cham: Springer.

DeWalt, B. 1994. Using indigenous knowledge to improve agriculture and natural resource management. *Human Organization* 53:123–131.

Duffy, R. 2000. *Killing for Conservation: Wildlife Policy in Zimbabwe*. Bloomington, IN: Indiana University Press.

Duruigbo, E. 2005. The World Bank, multinational oil corporations, and the resource curse in Africa. *University of Pennsylvania Journal of International Economics Law* 26:1–68

Dutta, S., A. Runacres and I. Sinha. 2018. Development-induced displacement, Indigenous knowledge, and the RFCTLARR act: A critical analysis. *Journal of Resources, Energy and Development* 15:25–37.

Eyong, C. T. 2007. Indigenous knowledge and sustainable development in Africa: Case study on Central Africa. *Indigenous Knowledge Systems and Development: Relevance for Africa* 1:121–139.

Fairhead, J., M. Leach and I. Scoones. 2012. Green grabbing: A new appropriation of nature? *Journal of Peasant Studies* 39:237–261.

Finer, M., C. N. Jenkins, S. L. Pimm, B. Keane and C. Ross. 2008. Oil and gas projects in the western Amazon: Threats to wilderness, biodiversity, and indigenous peoples. *PloS One* 3:e2932.

Ford, J. D., N. King, E. K. Galappaththi, T. Pearce, G. McDowell and S. L. Harper. 2020. The resilience of Indigenous peoples to environmental change. *One Earth* 2:532–543.

Gadgil, M., F. Berkes and C. Folke. 1993. Indigenous knowledge for biodiversity conservation. *AMBIO* 151–156.

Galtung, J. 1969. Violence, peace, and peace research. *Journal of Peace Research* 6:167–191.

Global Witness. 2020. Defending tomorrow. *Global Witness*. https://www.globalwitness.org/en/campaigns/environmental-activists/defending-tomorrow/?msclkid=e8c11533adbd11ec9ca374b70eaf4862

Greco, E. 2020. Africa, extractivism and the crisis this time. *Review of African Political Economy* 47:511–521.

Green, E. C. 1996. Indigenous knowledge systems and health promotion in Mozambique. In *Indigenous Knowledge Systems and its Uses in Southern Africa*, ed. H. Normann, I. Snyman and M. Cohen, 51–65. Pretoria: HSRC.

Green, L. 2013. *Contested Ecologies: Dialogues in the South on Nature and Knowledge*. Pretoria: HSRC.

Hallowes, D. and V. Munnick. 2016. *The Destruction of the Highveld Part 1: Digging Coal*. Pietermaritzburg, South Africa: groundWork.

Hallowes, D. and V. Munnick. 2018. *Boom and Bust in the Waterberg: A History of Coal Mega Projects*. Pietermaritzburg, South Africa: groundWork.

Harvey, D. 2005. *The New Imperialism*. Oxford: Oxford University Press.

Hens, L. 2006. Indigenous knowledge and biodiversity conservation and management in Ghana. *Journal of Human Ecology* 20:21–30.

Hilson, G. 2002. Harvesting mineral riches: 1000 years of gold mining in Ghana. *Resources Policy* 28:13–26.

Hiner, C. C. 2016. Divergent perspectives and contested ecologies: Three cases of land-use change in Calaveras County, California. In *A Comparative Political*

Ecology of Exurbia, ed. L. E. Taylor and P. T. Hurley, 51–82. Cham: Springer International Publishing.

Holifield, R. 2001. Defining environmental justice and environmental racism. *Urban Geography* 22:78–90.

Holm, P., J. Adamson, H. Huang, L. Kirdan, S. Kitch, I. McCalman, J. Ogude et al. 2015. Humanities for the environment. A manifesto for research and action. *Humanities* 4:977–992.

Homer-Dixon, T. F. 1994. Environmental scarcities and violent conflict: Evidence from cases. *International Security* 19:5–40.

Iheka, C. 2018. *Naturalizing Africa: Ecological Violence, Agency, and Postcolonial Resistance in African Literature*. Cambridge: Cambridge University Press.

Islam, N. and J. Winkel. 2017. Climate change and social inequality. *DESA Working Paper No.* 152.

Jenkins, K. 2018. Setting energy justice apart from the crowd: Lessons from environmental and climate justice. *Energy Research & Social Science* 39:117–121.

Karl, T. L. 1997. *The Paradox of Plenty: Oil Booms and Petro-states*. Berkeley, CA: University of California Press.

Katsaura, O. 2010. Violence and the political economy of informal diamond mining in Chiadzwa, Zimbabwe. *Journal of Sustainable Development in Africa* 12:340–353.

Klaaren, J. 2021. Xenophobia-induced disaster displacement in Gauteng, South Africa: A climate change litigation perspective. *Carbon & Climate Law Review* 15:150–157.

Kwashirai, V. 2009. *Green Colonialism in Zimbabwe, 1890–1980*. New York: Cambria Press.

Larmer, M. and V. Laterza. 2017. Contested wealth: Social and political mobilisation in extractive communities in Africa. *The Extractive Industries and Society* 4:701–706.

Larteza, V. and J. Sharp. 2017. Extraction and beyond: People's economic responses to restructuring in southern and central Africa. *Review of African Political Economy* 44:173–188.

Legesse, A., B. Teferi and A. Baudouin. 2013. Indigenous agroforestry knowledge transmission and young people's participation in agroforestry practices: The case of Wonago Woreda, Gedeo Zone, Southern Ethiopia. *Acta Geographica-Trondheim: Serie A* (26).

Lipper, L., P. Thornton, B. M. Campbell, T. Baedeker, A. Braimoh, M. Bwalya, P. Caron et al. 2014. Climate-smart agriculture for food security. *Nature Climate Change* 4:1068–1072.

Masolo, D. A. 1994. *African Philosophy in Search of Identity*. Bloomington, IN: Indiana University Press.

Mavhura, E. and S. Mushure. 2019. Forest and wildlife resource-conservation efforts based on indigenous knowledge: The case of Nharira community in Chikomba district, Zimbabwe. *Forest Policy and Economics* 105:83–90.

McGregor, D., S. Whitaker and M. Sritharan. 2020. Indigenous environmental justice and sustainability. *Current Opinion in Environmental Sustainability* 43:35–40.

McKay, B. M. 2020. Food sovereignty and neo-extractivism: Limits and possibilities of an alternative development model. *Globalizations* 17:1386–1404.

McQuade, J. 2019. Earth Day colonialism's role in the overexploitation of natural resources. *https://theconversation.com/earth-day-colonialisms-role-in-the-overexploitation-of-natural-resources-113995* (accessed January 16, 2021).

Mercer, J., I. Kelman, L. Taranis and S. Suchet-Pearson. 2010. Framework for integrating indigenous and scientific knowledge for disaster risk reduction. *Disasters* 34:214–239.

Mistry, J. and A. Berardi. 2016. Bridging indigenous and scientific knowledge. *Science* 352:1274–1275.

Mitman, G. 2017. Forgotten paths of empire: Ecology, disease, and commerce in the making of Liberia's plantation economy: President's address. *Environmental History* 22:1–22.

Moahi, K. H. 2007. Globalization, knowledge economy and the implication for indigenous knowledge. *The International Review of Information Ethics* 7:55–62.

Moore, J. W. 2017. The Capitalocene, part I: On the nature and origins of our ecological crisis. *The Journal of Peasant Studies* 44:594–630.

Moore, J. W. 2018. The Capitalocene, part II: Accumulation by appropriation and the centrality of unpaid work/energy. *The Journal of Peasant Studies* 45:237–279.

Moyo, S. 2003. *The Land Question in Africa: Research Perspectives and Questions.* Senegal: Codesria.

Mundy, P. 1993. Indigenous knowledge and communication: Current approaches. Submitted to *Journal of the Society for International Development.*

Mwangi, O. 2007. Hydropolitics, ecocide and human security in Lesotho: A case study of the Lesotho Highlands Water project. *Journal of Southern African Studies* 33:3–17.

Ndebele, N. 1998. Game lodges and leisure colonialism. In *Blank: Interrogating Architecture After Apartheid*, ed. J. Hilton and I. Vladislavic, 10–14. Cape Town: David Phillips.

Neimanis, A. et al. 2015. Four problems, four directions for environmental humanities: Toward critical posthumanities for the Anthropocene. *Ethics & the Environment* 20 (1):67–97.

Neumann, R. P. 2004. Moral and discursive geographies in the war for biodiversity in Africa. *Political Geography* 23:813–837.

News24. 2020. Why the murder of a KZN environmental activist is worrying for SAHRC. *www.news24.com/news24/southafrica/news/why-the-murder-of-a-kzn-environmental-activist-is-worrying-for-the-sahrc-20201104* (accessed December 4, 2020).

Nixon, R. 2007. Slow violence, gender, and the environmentalism of the poor. *Journal of Commonwealth and Postcolonial Studies* 13:3–12.

Nygren, A. 2004. Contested lands and incompatible images: The political ecology of struggles over resources in Nicaragua's Indio-Maíz reserve. *Society and Natural Resources* 17:189–205.

Nyiwul, L. 2021. Climate change adaptation and inequality in Africa: Case of water, energy and food insecurity. *Journal of Cleaner Production* 278:1–11.

Obeng-Odoom, F. 2020. COVID-19, inequality, and social stratification in Africa. *African Review of Economics and Finance* 12:3–37.

Obi, C. 2007. Oil and development in Africa: Some lessons from the oil factor in Nigeria for the Sudan. In *Oil Development in Africa: Lessons for Sudan after the Comprehensive Peace Agreement.* Copenhagen: Danish Institute for International Studies Report no. 8.

Obi, C. 2014. Oil and conflict in Nigeria's Niger Delta region: Between the barrel and the trigger. *The Extractive Industries and Society* 1:147–153.

Ogude, J. 1999. *Ngugi's Novels and African History: Narrating the Nation.* London: Pluto Press.

Puaschunder, J. 2020. Mapping climate justice. In *Governance & Climate Justice*, 23–38. Cham: Palgrave Macmillan.
Rall, K. 2020. Environmentalists are under threat for defending their rights. *www.businesslive.co.za/bd/opinion/environmentalists-are-under-threat-for-defending-their-rights/* (accessed December 6, 2020).
Robin, L. 2018. Environmental humanities and climate change: Understanding humans geologically and other life forms ethically. *Wiley Interdisciplinary Reviews: Climate Change* 9 (1):1–18.
Robinson, W. C. 2003. *Risks and Rights: The Causes, Consequences, and Challenges of Development-induced Displacement*. An Occasional Paper. Washington, DC: The Brookings Institution-SAIS Project on Displacement.
Rose, D. B. et al. 2012. Thinking through the environment, unsettling the humanities. *Environmental Humanities* 1 (1):1–5.
Ross, M. L. 1999. The political economy of the resource curse. *World Politics* 51:297–322.
Schlosberg, D. 2013. Theorising environmental justice: The expanding sphere of a discourse. *Environmental Politics* 22:37–55.
Singh, K. 2020. Anti-mining activist gunned down in her KZN home. *www.news24.com/news24/southafrica/news/anti-mining-activist-gunned-down-in-her-kzn-home-20201023* (accessed December 4, 2020).
Sörlin, S. 2012. Environmental humanities: Why should biologists interested in the environment take the humanities seriously? *BioScience* 62 (9):788–789.
Sovacool, B. K. and M. H. Dworkin. 2015. Energy justice: Conceptual insights and practical applications. *Applied Energy* 142:435–444.
Steyn, L. 2020. Tendele mine expansion on the line over community opposition. *www.businesslive.co.za/bd/national/2020-11-18-tendele-mine-expansion-on-the-line-over-community-opposition/* (accessed December 6, 2020).
Takeuchi, S. 2022. *African Land Reform Under Economic Liberalisation: States, Chiefs and Rural Communities*. Singapore: Springer Nature Pte Ltd.
Terminski, B. 2013. *Development-induced Displacement and Resettlement: Theoretical Frameworks and Current Challenges*. Geneva, Switzerland: University of Geneva.
Thompson, H. E., L. Berrang-Ford and J. D. Ford. 2010. Climate change and food security in sub-Saharan Africa: A systematic literature review. *Sustainability* 2:2719–2733.
Trainer, T. 1990. A rejection of the Brundtland Report. *IFDA Dossier* 77:71–84.
UNEP. undated. *Environment and Vulnerability: Emergency Perspectives*. Geneva, Switzerland: United Nations Environment Programme.
Visagie, J. and I. Turok. 2020. Rural–urban inequalities amplified by COVID-19: Evidence from South Africa. *Area Development and Policy* 6:50–62.
Warren, D. M. 1991. *Using Indigenous Knowledge in Agricultural Development*. Washington, DC: The World Bank.
Watts, M. J. 1999. *Petro-violence: Some Thoughts on Community, Extraction, and Political Ecology*. Berkeley workshop on Environmental Politics. University of California, 24–26 September.
West, P., J. Igoe and D. Brockington. 2006. Parks and peoples: The social impact of protected areas. *The Annual Review of Anthropology* 35:251–277.
Whyte, K. 2020. Too late for indigenous climate justice: Ecological and relational tipping points. *Wiley Interdisciplinary Reviews: Climate Change* 11: e603.

Williams, J. 2012. The impact of climate change on indigenous people – the implications for the cultural, spiritual, economic and legal rights of indigenous people. *The International Journal of Human Rights* 16:648–688.

Wood, N. and K. Roelich. 2019. Tensions, capabilities, and justice in climate change mitigation of fossil fuels. *Energy Research & Social Science* 52:114–122.

World Meteorological Organisation. 2020. *State of the Climate in Africa 2019*. Geneva, Switzerland: World Meteorological Organisation.

Youens, K. 2020. 22 October. *https://twitter.com/kirstenyouens* (accessed December 4, 2020).

Zalik, A. 2004. The Niger Delta: 'Petro violence' and 'partnership development'. *Review of African Political Economy* 31:401–424.

2 Petitioning the Future Through Environmental Justice. A Reading From Angola[1]

Catarina Antunes Gomes

Introduction

As an emerging and interdisciplinary field, the major contribution of Environmental Humanities may be described as the questioning of the ontological exceptionality of the human (Nye et al. 2013; Heise et al. 2017). This amounts to a development of the critique against anthropocentrism, which can be traced in the mainstream academia in western philosophy to Teilhard de Chardin and, more recently, to the 1960–70s proposals on Deep Ecology. Deeply informed by a critical reading of the science of his time, the philosophy of Teilhard de Chardin (Rideau 1965) emphasises the embeddedness of humanity in the universe and re-introduces an ethic of planetary care, conveyed by the recovery of the notion of the earth as Gaia.

In its turn, systematised by Arne Naess, the deep ecology movement is anchored on ethics respecting nature and the worth of non-human beings and aims to question and transform the root causes of environmental imbalances. Thus, deep ecology supersedes what is named commonly as "a shallow approach to ecology" in the sense that it refuses to adopt consumption or market-oriented values and methods which are foundational to our contemporary world. This refusal begets the questioning of our prevailing philosophies of development, while at the same time re-qualifying "ecosophies" and local systems of knowledge and cosmovision which assert the embeddedness of human and nature worlds, abolishing consequently dichotomic approaches to life. The 1992 Earth Charter represented another important consolidation of that same critique. Approved in 2000 at UNESCO in Paris, the Charter poses us two main propositions. These can be summarised as follows.

One is the recognition that the planet is one life-bearing community that is facing grave threats and that consequently, we all should adopt an *ethos* of co-responsibility – one that Leonardo Boff designated ethics of care. For the Brazilian author, "Care is, in truth, the real support of creativity, freedom and intelligence. In care, one finds the human *ethos*. . . . Care serves as a critique to our agonizing civilization and also as an inspirational principle for a new paradigm of convivialism" (1999: 2–3).

DOI: 10.4324/9781003287933-2

The second proposition that I would like to emphasise is the critique to the prevailing modes of development and notions of progress as material/industrial growth. According once again to the Earth Charter:

> The dominant patterns of production and consumption are causing environmental devastation. . . . Communities are being undermined. The benefits of development are not shared equitably. . . . Injustice, poverty, ignorance, and violent conflict are widespread and the cause of great suffering.

Therefore, the text postulates that Fundamental changes are needed in our values, institutions, and ways of living. We must realise that when basic needs have been met, human development is primarily about being more, not having more. In these examples and references, in which the Paris Agreement is inscribed, what we find is a fundamental appeal to overcome the self-referentiality of the dominant framework regarding the place of the human, the place of nature, and related models of development.

That dominant framework is built upon a reifying cosmovision that has placed historically the human at and as the centre of existence. We all know the general roots and logical corollaries of such a claim. The human becomes the demiurge with the right to dispose, use and transform existence. This has been named the sin of anthropocentrism or "the God Complex." To quote again Boff:

> The modern human being created a God Complex. . . . Through the technoscience project, she/he believed that she/he could accomplish everything and that there were no limits to her/his will to know, to dominate, to create. That pretense placed huge demands on her/himself. She/he can no longer stand the development that has already shown its destructive character by threatening the common fate of earth and its inhabitants.
>
> (1999: 7)

A meditation on these issues clearly forces us to face a demand for transformation – a transformation that can only live up to expectations if it is projected and lived as a transcendence effort to overcome the self-referentiality that still captures imagination, action, and our very own future – a self-referentiality that can only be aptly described as a possession project – one that is contrary to the very idea of liberation as the opening of a new horizon of being. And no transcendence project can become viable without a conscious commitment to and for care beyond possession. Such a commitment is a decision born out of struggle and choice. It is a decision of consciousness seeking to realise itself as a being-amongst-beings. A decision to sustain and to bear life into the future.

The Human That We Don't Expect

The historical celebration of anthropocentrism signalled the liberating effect of the human, through which she/he ascended to the status of a creator. However, this reconceptualisation of the human from creature into creator fed social and cultural imaginaries that soon came to legitimise the intensification of a reifying gaze upon existence. Anthropocentrism became the signifier of the demiurge's right to possess and dominate. Foucault (1997) traced the complex relationships between what can be designated as a will for knowledge, a will for power and a will for truth in critically assessing the historical constitution of science, politics, and ethics and their role in shaping the prevailing western globalised cosmovision. It is perhaps noteworthy to also recall Sartre's critique concerning the self-referentiality of any act of knowledge, which was formulated in terms of the Acteaon complex. In his view, knowledge signifies a possession that ultimately leads to assimilation as dissolution – the ultimate act of violence:

> Every investigation implies the idea of a nudity which one brings out into the open by clearing away the obstacles which cover it, just as Actaeon clears away the branches so that he can have a better view of Diana at her bath. More than this, knowledge is a hunt. . . . The scientist is the hunter who surprises a white nudity and who violates it by looking at it. . . . To know is to devour with the eyes.
> (Sartre 1984: 738–739)

Hence, "there is a movement of dissolution which passes from the object to the knowing subject. The known is transformed into me, it becomes my thought and thereby consents to receive its existence from me alone" (Sartre 1984: 739).

Can we move beyond this self-referentiality? Can we move beyond an *ethos* of possession? Ironically, anthropocentrism as a consecrated right of possession and domination has been inverted precisely by placing once again the human as the engine of planetary change. The inversion designates a new description of the human: not as a rightful demiurge but as the cause of climatic change. And, this time, as the engine of planetary change, the human is called upon and issues of responsibility in terms of life sustainability come to the forefront. The recognition of humanity's role in changing dramatically the conditions to sustain and bear life demands the critical questioning of the human's right to its own supposed "ontological exceptionality."

We are now left in the Anthropocene. As stated, and in a seemingly paradoxical way, an Anthropocene that questions precisely anthropocentrism. Introduced by Paul Crutzer and Eugene Stoemer in the early 2000s, Anthropocene designates briefly the humanity as a planetary geological concept, "a post-Holocene epoch defined by the significant impact of humans on geological, biotic and climatic planetary processes" (Neimanis et al. 2015: 2).

And as explained by Ellis, "the significance of the Anthropocene resides in its role as a new lens through which age-old narratives and philosophical questions are being revisited and re-written . . . with the potential to radically revise the way we think of what it means to be human" (2018, Kindle edition). DeLoughney also points out crudely the following: "Understanding climate change as a geological shift created by humans leads to new conceptions of history, deep time, and of the notion of humanity, which in turn raise important questions in considering scales of ontology" (2015: 12).

So instead of postulating the "ontological exceptionality of human" translated historically in a language of power and a practice of possession, the concept of Anthropocene allows us to re-think that exceptionality, as well as the relationship between human and non-human existence. In this sense, the concept represents another step in the systematisation of the critique of global capitalism, embedded in colonial frameworks, and an urgent appeal to not succumb to its autopoietic self-referentiality.

But a silent question remains: who is human? Or to put it differently: who has been exceptional enough to be considered as such?

Arising from different fields and concerns, such as political ecology, literary ecocriticism, and post-colonialism, that silent question becomes articulated in an endless procession of negativities: the narratives and experiences of those that have been in so many ways dispossessed – dispossessed even of the status of being "human." Phenomenology becomes here the agent of denouncing the shortcomings and pitfalls of the human demiurge and history as the stage of the de-realisation of the human. And how is it that the historical ontological exceptionality of the human leads ultimately to its de-realisation?

The concept of Anthropocene allows us to re-think anthropocentrism as a force of what Butler named "human derealization" Palsson et al. (2013) see the concept as designating a "new human condition." In its dystopian promises, this human condition is one of impossibility to which we are led through violent forms of de-realisation of the human. For Butler:

> I am referring not only to humans not regarded as humans, and thus to a restrictive conception of the human that is based upon their exclusion. It is not a matter of a simple entry of the excluded into an established ontology, but an insurrection at the level of ontology, a critical opening up to questions, What is real?, Whose lives are real?, How might reality be remade? Those who are unreal have, in a sense, already suffered the violence of derealization. . . . Does violence take place on the conditions of that unreality? . . . The derealization of the 'Other' means that it is neither alive nor dead, but interminably spectral.
>
> (2006: 33–34)

In order to address these issues, it might be useful to understand how the challenge to the ontological exceptionality of the human should not be made in an historical void, whereby historical and political responsibilities would be nullified. It is in this sense that a post-colonial approach becomes crucial: it is the element needed in order to understand how beings and realities are made unreal, turned into spectres, and imagine and implement new ways of life, new scopes of ontology, and new realms of possibility.

One major contribution of a post-colonial approach in environmental humanities is the awareness that the history of colonialism and capitalism are foundational dynamics for the modern celebration of the human demiurge and for the reification of human and non-human existence, drawing (in a seemingly perpetual rhythm) the Fanonian lines between zones of being and non-being. That same awareness might help us to understand that there are very different ecological agencies and responsibilities as well as very different answers to the challenge of questioning the "ontological exceptionality of the human."

There is clearly a conceptual tension between human agency as a species, encapsulated by the concept of Anthropocene, and the inequalities that shape the ecological agency of different communities. For example, as stated by Paulsson et al., "the notion of 'ecological debt' and 'climate debt' have been used to try to capture how the global North's excessive historic use of the atmosphere's absorptivity capacity has closed off similar development routes for the global South. . . . The very language and metric used in climate policies and negotiations can mask such issues" (2013: 7). In fact, from the spectral point of view of human derealisation, what does it mean the questioning of human ontological exceptionality?

The Spectres' Whispers

As it is known, Angola was under Portuguese colonial rule until the mid-1970s. The last century was prolific in colonial politics of redesigning landscapes, peoples, and resources while implementing capitalist modes of extraction and economic development. In fact, as we know, the Berlin Conference (1884–85) stipulated that the "legitimate ownership" of the colonies depended upon an effective occupation of those territories. In this context, Portugal faced great pressure to prove its effective presence in the colonies. That led to the beginning of occupation politics, especially in Angola and Mozambique. And that occupation politics needed to be not only demographic, but also economic.

In the late XIX century and early XX century, Angola and Mozambique were mainly seen as exiles for forced and convicted settlers. However, in Angola, settler policies were deployed especially in the 1920s by High Commissioner Norton de Matos, known as the "steel surgeon," through the development of colonial administration and the migration of Portuguese peasants and unemployed workers. Portuguese colonies became the synonym of the expansion of Portugal. For sure, the effective settler-occupation needed to be anchored on modes of capitalist investment and development. Infrastructures were built to support commerce, companies with foreign capital were given access to natural resources, while cheap forced labour represented another crucial competitive advantage.

The colonies' economic development motivated and facilitated Portuguese migration, which increased steadily after the Second World War till the beginning of the liberation wars in the 1960s. Portuguese colonial rule paid special attention to the development of capitalist agricultural projects. In Angola, two major projects were developed: the Cela settler project and the Maiata settler project. The diamond mining was another example: Diamang was an international capitalist enterprise created in 1917 that operated as a state within the colonial state, endowed with the direct power to fully regulate all aspects of a forced labour force's life.

Settler projects had a special violent character. They were implemented to the detriment of local populations through land usurpation and expropriation. Local communities were displaced to indigenous reserves or even concentration camps. One example was the São Nicolau Imprisonment Camp in Namibe Province. Freeing the land for European occupation and exploitation, these strategies also made available for colonial authorities' sufficient labour force for the capitalist exploitation of the land and its resources. Disarticulating local communities and their own systems of knowledge, these policies amounted to an alienation strategy that was further aggravated by the introduction of differentiation policies between local populations, feeding a sense of rivalry and betrayal between them.

One such policy concerned the "black troops" of the Portuguese army. Especially during the liberation struggle, the incorporation of black soldiers in the Portuguese army produced, to this day, important fractures in the social fabric. Their experience can perhaps be related to the very well-known experience of *SonderKommands* in Nazi concentration camps, portraited by Primo Levi, whereby subjects were submitted to and lived in situations of moral impossibility. These examples affected particularly the San communities in Angola. San communities live in the southern provinces of Huila, Cunene, and Kuando-Kubango. They were classified by colonial authorities as a non-bantu people, such as Ovakwissi, Ovalwepe, and N'Kung! The term Khoisan in use today is a colonial invention created to refer to two different people that apparently shared morphologic and ethnolinguistic traits: the San and the Khoi-Khoi (Bam 2015).[2]

During the liberation war, San men were recruited by the Portuguese military to conduct field recognition and identify and denounce possible enemies in local communities. Given their knowledge of the territories, they were incorporated into the regiment of soldiers, called "Flechas" ("Arrows") and their formal integration in the colonial army dates back to 1967. The goal was to collect information, to translate information, and to support military operations. By 1974, in Angola, there were 22,000 flechas. This practice led to a deepening of the fracture between the San and Bantu communities that persist today. The relationship between the two communities is marked by mistrust and the crude marginalisation of the San people treated as "primitive," "ignorant," "lazy," and "dirty." San communities are marginalised from the citizenship realm and, as a consequence, their social and economic integration within the nation-state is very frail. And let us keep in mind that climate change affects particularly "nearly 80 percent of the world's poor that live in rural areas and typically rely on agriculture, forestry and fisheries for their own survival," such as the San (FAO 2019: v).

In a vein similar to colonial policies, the current land grabbing phenomena and the option of agro-business have further excluded San communities from their traditional territories. Current state policies are forcing them to adopt a sedentary lifestyle without fully integrating those communities within the scope of universal citizenship rights. As a consequence, their systems of knowledge are being eroded, something that weakens their empowering character, and the degree of unemployment, discrimination, poverty, and overall exclusion remains very high.

The consequences of these phenomena, namely in the environment, as well as the lack of consequent public policies to fulfil basic human rights, are currently being magnified by serious climate change, affecting particularly the historically disavowed communities. For a number of years now, the south of Angola, where San communities live, has been affected by a drought that has seriously diminished the capacity of these communities to survive. Food insecurity and diseases related to the lack of access to water are spreading and conflicts between San and Bantu for water are becoming quite common. In the study coordinated by UNICEF (2018) which focused on San communities located in the Province of Huíla, testimonies about increasing tensions between San and Bantu families over access to water were reported. It was also reported how low is their access to basic sanitation and basic public services in health and education and the overall experience of being in the lowest step of social life. And one should keep in mind that this crisis is happening in a context where, according to official data, only 44% of the national population has access to drinkable water and where only 32% of households have adequate sanitary installations. This is happening in a context where chronic undernutrition in children under 5 years old is staggering, as Table 2.1 shows.

This is happening in a context where public policies favour industrial agriculture in the detriment of familial agriculture, which is the main source of food for the majority of the population. This is happening in a context where only 1.8% of the national budget is invested in water and sanitation and where recurrent epidemics of yellow fever and cholera are taking place (ADRA et al. 2018). And it is really worth mentioning that, despite the colonial category of "San people" being still in use in public discourses regarding poverty and vulnerability, there are no available stats to see how those indicators translate into and shape San's everyday life. They can be used as visible signposts for vulnerability, but they remain invisible in most diagnostics and public policies.

Table 2.1 Chronic undernutrition by Angolan Provinces – Children under 5 years old (IIMS 2015–2016)

Very High Prevalence Rate	Uíge	42%
	Bengo	40%
	Cuanza Norte	45%
	Cuanza Sul	49%
	Huambo	44%
	Cuando Cubango	43%
	Bié	51%
	Lunda Sul	42%
	Huíla	44%
High Prevalence Rate	Luanda	30%
	Benguela	33%
	Malanje	32%
	Lunda Norte	39%
	Moxico	39%
	Namibe	39%
	Cunene	39%
Medium Prevalence Rate	Zaire	25%
	Cabinda	22%

Source: Instituto Nacional de Estatística. *Inquérito de Indicadores Múltiplos em Saúde 2015–2016*. Luanda, Angola. Adapted by the author.

Nonetheless, the state insists on following capitalist-oriented policies. Large landowners are spreading throughout the territory in an attempt to promote industrial agricultural production. If in the colonial era, this was seen as the economic development of the colonial power (not necessarily the colony in itself), today these same policies are advocated for the cause of diversifying an economy that is almost completely oil-dependent. However, a number of paradoxes emerge once again. One, based on monocultures, this option entails once again expropriation and the privatisation of land and rivers as well as the displacement of communities and the impossibility of those communities to survive for themselves. Two, it is also clear that agriculture is the main source of food for the majority of the population. Industrialised agricultural projects aim especially at exportation rates, not to mention environmental consequences if left unregulated.

In the case of San communities, once again, access to rivers and water is often forbidden, as well as hunting for food. The active production of this new dimension of exclusion reinforces negative stereotypes, which amounts to shaming and blaming politics: "these communities are unable to evolve and integrate modernity. The state cannot be blamed by their ineptitude. After all, they are stubborn primitives" – comes to be the eloquently silent official argument. "They need to be resilient" – it's the corollary of current policy mottos. And let us be clear: resilience here is nothing more than an increased capacity to suffer since it does not question the structures and practices that are the cause of exclusion.

In public and official discourse, populations like the San, are described as "vulnerable." This sanitised and apolitical classification allows for these communities' reification in the politics of representation. As a technical category, it also allows for dismissing the active agency of the state in producing this disciplinary exclusion. This category – the vulnerable – is included in the apparatus of what Foucault named as governmentality, that is, the way in which political power controls and regulates populations and bodies. The regulatory power of governmentality (colonial and post-colonial) produces a set of authorised lives and authorised or tolerable deaths. The latter result arises from a process of de-subjectivation of a subject and leads to an overall derealisation of the human.

Unlike South Africa and Botswana, for example, where one finds civic organisations and social movements to defend San communities' rights, in Angola San's marginalisation is still aggravated by the lack of (real, not only formal) recognition of their human and citizenship status. Their relation to the state faces the challenge of assimilation or natural death caused by society's natural evolution and progress. They claim their human status and revolt for having to drink the dirty water reserved for dogs and cattle. In these cases, as Butler pointed out, no public grieving or mourning will take place.

On the other hand, and at the same time that trying to invest in industrial agriculture, the state persists in adopting emergency policies to address epidemic outbreaks instead of implementing low-cost strategies to guarantee access to water and basic sanitation – many of them based on local knowledge or with low cost, such as sand dams (ADRA et al. 2018).

Figure 2.1 Cacimba (handmade community non-drinkable water source for human and animal consumption and agriculture), Tchiquaqueia. Huíla.

Source: Photo by the author. 2017.

The irrationality of state policy can only be explained by the stronghold that powerful economic interests exercise in what concerns political decisions. It is more immediately lucrative to promote large landowners than local and communal agriculture, which is responsible, in fact, for 95% of Angolan food production, jeopardising food security (ADRA et al. 2018). It is more lucrative to establish contracts with agencies and companies for the construction of big infrastructures that tend to collapse due to the lack of maintenance and that do not serve more remote or rural villages than to invest in local sanitation or in the democratisation of health and education.

Thus, even with celebrated Earth Charters and global agreements, even with Sustainable Development Goals, what we encounter is not a transformational engagement with our world, but the intensification of what has been dominating our collective and individual lives: the intensification of capitalist modes of production that leads to the "de-realization of the human."

Just to give a few examples: the new dynamic of land grabbing observed in Africa has been legitimised and presented as "the green revolution of the continent" and also as the most suitable response to food security concerns. But, in fact, this "green revolution" is being led by multi-nationals that exhaust local resources, reduce the capacity of natural resources to sustain food demand, expropriate local communities, and contribute nothing to food security since their aim is the exportation of goods. Two-thirds of the private investment in agriculture are concentrated in countries with high rates of undernutrition (Seles 2019; Fernandes 2019). FAO has recently recognised that hunger is on the rise again since 2016. In Sudan, 9.8 million acres[3] were sold to foreign and private corporations; in Mozambique 6.6 million and in Liberia 4 million (Seles 2019). Is this really a "green revolution"? For whom?

Perhaps we should recall that Eric Toussaint has explained to us that land privatisation is a key element of structural adjustment programmes for the so-called developing countries and that it is a strategy needed for these countries to pay their increasing debts. A strategy that also erodes local ecosophies and that makes communities into dispossessed communities of their own knowledges: "Over the centuries, Third World peasants have been the ones to produce . . . the very wealth of the soil that multi-nationals are now stealing from them. As though it were not enough for big agro-business firms to make off with the genetic maps of these agricultural organisms, they also patent them, and demand royalties be paid in exchange for use" (1999: 248).

There is a sort of an enlightening genealogy in predatory capitalism and its orthodoxies. One proof of such a remarkable ability is the fact that it survived crucial political moments of redefining humanity's project. Its emergence and consolidation through the colony have been sophisticated with and in the post-colony. Humanity's project becomes often and cynically used as a poetic device to pave the way for global capitalism's legitimacy so that no structural decisions can be made. The full transubstantiation of domination into liberation implies certainly that *a luta continua*.[4]

The point here is that this intensification process of capitalist appropriation of land and radical climate change not only endangers the right to be but also poses a radical question: where and how is one to be? We have been witnessing increasing flows of different kinds of migration, such as "out-migration," "forced displacement," and "survival migration."[5] According to FAO (2018), more than 1 billion people living in developing countries have moved internally as part of current economic transformations. Many of these migrants are refugees or internally displaced people. And it should be clear to all of us that internally displaced people have no legal status and thus may not claim any additional rights to those shared by their co-citizens. And also that although the category of the refugee has a legal status, it does not encompass migration due to natural disasters, abrupt climate changes, and poverty. FAO estimates that approximately 18 million people were displaced between 2008 and 2017 due to climate change and that the "likelihood of being displaced by a disaster increased by 60 percent from 1970 to 2014 and it is expected to continue growing" (2018: 45). The impact of all this on Angolan San communities remains to be fully unveiled. Beyond being vulnerable and in need of resilience, are they to be seen as "internally displaced"? Metaphorical refugees?

On the other hand, according to "The Millennium Development Report 2015" – a document that informed the Sustainable Development Goals,

> In 2011, 41 countries experienced water stress, up from 36 in 1998. Of these, 10 countries – from the Arabian Peninsula, Northern Africa and Central Asia – withdrew more than 100 per cent of renewable freshwater resources. Once a country reaches a withdrawal level above 100 percent, it starts depleting its renewable groundwater resources, relying on non-renewable fossil groundwater or non-conventional sources of water, such as desalinated water, wastewater and agricultural drainage water. Currently, water scarcity affects more than 40 percent of people around the world, and it is projected to increase. Water scarcity already affects every continent and hinders the sustainability of natural resources as well as economic and social development.
> (UN 2015: 55)[6]

What is at stake here is the self-referentiality of the development model in use. It has been no coincidence that the Anthropocene is related to the Capitalocene. The agent of planetary change was set loose by predatory capitalism. Authors like Escobar have taken a step forward in the mainstream approach to development and talk about "post-development" (1992). Besides capitalism's autopoietic nature, two fallacies continue to reproduce themselves: first, the dichotomy between human and nature and, second, the very question of the "human." In fact, for instance, San communities seem to escape this last category or not to fully embody the idea of human. Neither in life nor in death, they need to be managed almost the same way nature needs to be managed. It is in this precise sense that San communities continue to be the object of de-realisation of the human. So, who is the human? Who is the human that needs to be human and the human that needs to abandon its ontological exceptionalism and ex-position itself as nature? And what does it mean when San communities, as others, claim their status as humans?

Who Is Real Without a Future? Or Love as a Decision to Sustain and Bear Life

For Butler, "when we recognize another, or when we ask for recognition for ourselves, we are not asking for an Other to see us as we are, as we already are, as we have always been . . . .. To ask for recognition, or to offer it, is precisely not to ask for recognition for what one already is. It is to solicit a becoming, to instigate a transformation, *to petition for the future*. . . . It is also to stake one's own being, and one's own persistence in one's own being" (Butler 2006: 44).

The point made by Butler is crucial: an existence without a prospect for a future is a death in life. Having a future is a privilege. It is what certifies us as worthwhile beings. A politics of recognition is not to be anchored in emergency situations; It is essentially a background work to sustain and be able to bear life into the future. It is a work of transformation of our current and prevailing cosmovision and practices in order to counteract not only environmental predation but also human derealisation.

The impact of the combination between global capitalism and ecological collapse upon what we now term as the global south is dreadful. It is easy to picture how an age of human derealisation will most probably entail a human rarefaction.

More than 60 years ago, Teilhard de Chardin wrote: "fifty years ago, we would imagine ourselves to look at the earth décor surrounding us as inactive and irresponsible spectators. We were children. Today we understand that we are workers connected through an immense task. We feel like the living atoms of a moving Universe. We became adults" (*apud* Rideau 1965: 36).

Toussaint posits himself between the pessimism of the intellect and the optimism of the will. And perhaps we are indeed adults. What decisions are we to make? Will we petition for the future?

> *Father, why do you look so much into the past?*
> *Because in the past there was future.*[7]

Postscript[8]

To petition the future.

To petition the future is assuring that the fundamental possibility of becoming is granted.

In the midst of the C-19 global pandemic and in the anticipation of a new radical long-term crisis, to petition the future is to reassess and reimagine this human factor that has been producing so avidly planetary change. One cannot sustain an "innocent gaze" after the massive and successive warnings.

The last Global Health Security Index (GHS Index)[9] was published in December 2019. Wary of (natural, intentional and accidental) biological threats that represent great risks for global health and humanity, the report classified 195 countries in terms of their readiness to face such threats (involving prevention, detection and reporting, emergency response, health system and compliance with international norms. The main finding is that the average of all 195 countries is 40.2 points (0–100 points), a remarkable (or expected?) low score. The report explains the result in terms of prevention, detection and reporting, emergency response, health systems, environmental risks, etc:

> Overall, the GHS Index finds severe weaknesses in country abilities to prevent, detect, and respond to health emergencies; severe gaps in health systems; vulnerabilities to political, socioeconomic, and environmental risks that can confound outbreak preparedness and response; and a lack of adherence to international norms.
>
> (2019: 9)

Hence, the main conclusion:

> Countries are not prepared for a globally catastrophic biological event, including those that could be caused by the international spread of a new or emerging pathogen or by the deliberate or accidental release of a dangerous or engineered agent or organism. Biosecurity and biosafety are under-prioritized areas of health security, and the connections between health and security-sector actors for outbreak response are weak.
>
> (2019: 10)[10]

The un-incredible plausibility of an "innocent gaze" facing the pandemic reduces agency to dust and it reframes the very notion of free will within a new framework: fatalism, a barren imagination for conscious free will to prosper. Environmental Humanities represent a crucial effort to build a sustainable, life-friendly, and democratic worldliness – one able to petition the future.

But, again, to petition the future is a decision.

Notes

1 All quotations in foreign languages were translated into English by the author.
2 However, they are quite different from one another. For example, Khoi-Khoi used to work metal and their economy was based on cattle. The San, on the other hand, were known by being hunters-gatherers.
3 1 acre = 0.404 hectares.
4 The liberation motto, *A luta continua!* meaning 'The struggle goes on!'.

5 Out-migration is the movement of a person or group out of one community in order to reside in other; Forced displacement is the result of coercive factors arising from natural or human-made causes; Survival migration refers to people who perceive that there are no local options to survive with dignity (FAO 2018).
6 More recently, the 'Report on Household drinking water, sanitation and hygiene 2000–2017', published by WHO and UNICEF in 2019, stresses the following: "hygiene comprises a range of behaviours that help to maintain health and prevent the spread of diseases, including handwashing, menstrual hygiene management and food hygiene. The indicator selected for global monitoring of SDG 6.2 is the proportion of the population with a handwashing facility with soap and water available at home. In 2017, 60% of the global population (4.5 billion people) had a basic hand washing facility with soap and water available at home. A further 22% (1.6 billion people) had handwashing facilities that lacked water or soap at the time of the survey, and 18% (1.4 billion people) had no hand washing facility at all" (WHO and UNICEF 2019: 35).
7 Gomes, Catarina Antunes. The liars' book. Unpublished manuscript.
8 The first version of this paper was presented at the Colloquium on Environmental Humanities, held by the Centre for the Advancement of Scholarship, University of Pretoria, South Africa (May 2019), and it benefited deeply from such a gathering of concerned scholars. It was revised in May 2020, and this is the reason why a postscript was added.
9 Global Health Security Index. Building collective action and accountability. October 2019. NTI/ John Hopkins Bloomberg School of Public Health.
10 The report was criticized by missing some of their conclusions. For instance, the report stated that the US and the UK were the most well-prepared countries in the world. Nonetheless, in the Global North, they hit very hard. Surprising? Contradictions can also be evidence.

References

ADRA, UNICEF, and E. MOSAIKO. 2018. Folhetos temáticos do Orçamento Geral do Estado em Angola. *www.unicef.org/angola/relatorios/an%C3%A1lise-geral-do-or%C3%A7amento-geral-do-estado-2018* (accessed January 6, 2019).

Bam, J. 2015. Contemporary Khoisan identities in the western Cape and campaign for social justice. *Bulletin of the National Library of South Africa* 69:215–232.

Boff, L. 1999. *Saber Cuidar Ética do humano – compaixão pela terra*. Rio de Janeiro: Editora Vozes.

Butler, J. 2006. *Precarious Life. The Powers of Mourning and Violence*. London: Verso.

DeLouhney, E., E. J. Didur and A. Carrigan. 2015. *Global Ecologies and the Environmental Humanities. Postcolonial Approaches*. New York: Routledge.

Ellis, E. C. 2018. *Anthropocene. A Very Short Introduction*. Oxford: Oxford University Press.

Escobar, A. 1992. Imagining a post-development era? Critical thought, development and social movements. *Social Text* 31/32 (Third World and post-colonial issues):20–56.

FAO. 2018. *The State of Food and Agriculture. Migration, Agriculture and Rural Development*. Rome: FAO.

FAO. 2019. *Agriculture and Climate Change. Challenges and Opportunities at the Global and Local Level*. Rome: FAO.

Fernandes, A. 2019. A posse da terra, a recolonização silenciosa que não diz o nome *Africa* 21:22–25.
Foucault, M. 1997. Ethics: Subjectivity and truth. In *Essential Works of Michel Foucault 1954–1984*, ed. P. Rabinow. London: Penguin.
Global Health Security Index. 2019. *Building Collective Action and Accountability.* NIT/John Hopkins, Bloomberg School of Public Health. October. *https://www.ghsindex.org/wp-content/uploads/2020/04/2019*
Heise, U., J. Christensen and E. M. Nieman. 2017. *The Routledge Companion to the Environmental Humanities.* London: Routledge.
Instituto Nacional de Estatística. 2016. *Inquérito de Indicadores Múltiplos de Saúde 2015–2016.* Luanda: INE.
Mondlane, E. 1995. *Lutar por Moçambique.* Colecção 'Nosso Chão': Maputo.
Neimanis, A., C. Asberg and J. Hedrén. 2015. Four problems, four directions for environmental humanities: Toward critical posthumanities for the Anthropocene. *Ethics and the Environment* 20:67–97.
Nye, D., L. Rugg and J. Fleming. 2013. *The Emergence of Environmental Humanities.* Stockholm: Mistra, The Swedish Foundation for Strategic Environmental Research.
Palsson, G. et al. 2013. Reconceptualizing the 'anthropos' in the Anthropocene: Integrating the social sciences and humanities in global environmental change research. *Environmental Science & Policy* 28:3–13.
Rideau, É. 1965. *O pensamento de Teilhard de Chardin.* Lisboa: Morais Editora.
Sartre, J-P. 1984. *Being and Nothingness.* New York: Washington Square Press.
Seles, J. 2019. O dilema das terras em África. Na viragem do milénio. *Africa* 21:26–27.
Toussaint, E. 1999. *Your Money or Your Life! The Tyranny of Global Finance.* Virginia and Dar Es Sallam: Pluto Press and Mkuki na Nyota Publishers.
UN. 2015. *The Millennium Development Report.* New York: UN.
UNICEF. 2018. Saúde materno-infantil nos municípios da Huíla: conhecimento, atitudes e práticas. *www.unicef.org/angola/relatorios/saude-materno-infantil-nos-municipio* (accessed January 10, 2019).
WHO and UNICEF. 2019. Report on household drinking water, sanitation and hygiene 2000–2017. *https://www.who.int/publications/i/item/9789241516235?msclkid=f38cd90cadc211ecb5ac9f44419cfbfb*

3 African Goats, the State and Conservation in Colonial Zimbabwe, 1892–1970s

Elijah Doro

Introduction

In 1946, French historian Emille Segui (1946) wrote:

> The billy goats and their flocks are the most serious enemy of viticulture, woodland management, olive groves, and orchards and above all for all those who are not especially diligent, of the kitchen gardens. They are nasty, odious, bad tempered, noisy, beasts distinguished particularly by the stink of their bad breath. From the times of antiquity until our present era people have been of the same opinion

This invective apostrophised the anti-goat zeitgeist and capriphobia that was dominant in much of global conservation thinking during this time and had its roots in antiquity. Indeed, the goat (*Capri Hircus*) was one of the most maligned domestic animals and the bêta noire of most conservationists across history. From classical times, goats were viewed as an aberration to the normative livestock regimes and a scourge to the woodlands, pastures, and the environment (Virgil 1969).[1] Capriphobes often disparaged the goat for its voracious browsing of thick bush, grasslands, and frenetic trampling of the ground in steep lands and mountainous areas that precipitated deforestation and hastened land degradation. Goats were also implicated in the degeneration of insular ecosystems in Island colonies such as St Helen and Santa Catalina where deforestation, soil erosion, habitat destruction, and extinction of species had occurred from extensive vegetation removal.[2] Consequently, goats fell victim to proscriptions, environmental crusades, conservation pogroms, and bans across the world that marginalised them to the peripheral spaces in pasture landscapes. This anti-goat conservation movement had its roots in Europe and the Mediterranean world where increasing goat populations from the 15th century had been blamed for the destruction of forestry and land resources. In 1453, the Canton of Freiburg in Switzerland prohibited the grazing of goats and most Swiss cantons forbade the grazing of goats in coppice stands (Maher 1945). Goats were banned from grazing in forests in Israel in 1953 and in Morocco and Tunisia in 1959.

DOI: 10.4324/9781003287933-3

This wave of anti-goat conservation was popularised in Africa by colonial officials during the apogee of enforced land centralisation and conservation in the 1930s. While there was consensus that overstocking was causing land degradation in African areas, the blame was disproportionately apportioned to goats ahead of cattle and other stock. Goats were scapegoated for deforestation, rangeland destruction, soil erosion, and desertification. This culminated in the enactment of statutes prohibiting the grazing of goats in certain areas, pasturage fines, destocking, and goat eradication campaigns.

However, while the African goat bore much of the brunt of discrimination and environmental injustice from the colonial state, it has received less attention than cattle in the livestock historiography of southern Africa. Indeed, as (Badenhorst 2006: 45–53) asserts much of the existing literature on livestock in southern Africa has placed much emphasis on cattle obscuring the role of goats in both the precolonial and colonial periods. The "cattle complex" has captured the attention of many livestock historians who have used narratives of cattle keeping as lenses to view the colonial order (Mwatwara 2014; Samasuwo 2000). Historical studies that have looked at goats in southern Africa are mostly archaeological, focusing on occurrences of goats in rock art, their sizes, and distribution in precolonial iron age societies (Robinson 1986; Badenhorst 2002; Brink and Holt 1992). This chapter look at the colonial order of land centralisation and conservation from the point of view of African goats. In doing so, it interrogates the role of livestock in changing environmental landscapes during the era of the Anthropocene. It shows the political ecology of livestock regimes in colonial Africa and the power dynamics governing the allocation of space in pasture and social landscapes. It uses African goats as animals of the subalterns to engage with themes of environmental justice and environmental racism. The chapter unpacks how the discourses of anti-goat conservation in Africa were constructed along the lines of hygiene and race. Ultimately, the chapter frames goats and goat farming within broader contemporary debates on sustainability and climate change.

The "Poor Man's Cow": Goats, Subalterns, and Marginalisation in History

Goats were amongst the first species of animals to be domesticated around 10,000 BC in the fertile crescent of the Middle east (Dwyer 2017). Archaeological evidence points out that goats were present amongst iron age people in the eastern half of southern Africa (Badenhorst 2006). The ability of the goat to survive harsh environmental conditions while providing the means of subsistence (milk, meat, and skins) to farmers has led it to be called the "poor man's cow." Throughout history, goats have provided the most suitable and stable form of investment for poor families because of their adaptation to marginal areas and arid regions where cattle, sheep, and other domestic animals are less likely to thrive (Siddle 2009). However, by the

40 Elijah Doro

Middle Ages, a series of economic and political changes in Europe impinged on the status of the goat in the livestock hierarchy and relegated them to outcasts. The agricultural revolution and its attendant enclosure movement and the rise in commercial wool production played a crucial role in marginalising goats. The enclosure movement reduced common land that had guaranteed pasture for goats and witnessed a decline in the number of goats as they were now seen as a nuisance to crops (Dohner 2001). The emergence of commercial wool production placed greater value on sheep and pushed goats from the grazing lands into the woodlands. In the mountainous woodlands, the goat came into conflict with foresters and the aristocracy who needed timber to build ships and Hunting turf, respectively. Siddle (2009) argues that the earliest bans on goats sprang from this conflict between the poor man's goats on one side and the rich man's commercial and leisure interests rather than a genuine concern for the environment.

The polemicisation of the goat extends to its iconographic representation as a "scapegoat," the repository of sin and its demonisation in western and African culture.[3] The goat came to represent an ideology of command and control, the subversion of the weak by the strong. In ancient Europe, goats and goat keepers came to be reviled on account of their habitation of mountainous areas which were viewed as incubators of anarchy and civil strife (Siddle 2009). Goats were also anthropomorphised and came to symbolise the base qualities of lasciviousness, halitosis, and anarchic deviousness. The earliest goat bans were a way of controlling the anarchic elements of the population, preserving hunting grounds for the aristocracy and protecting the wool industry, and improving timber supplies for shipbuilding.

As European countries opened colonies in other parts of the world, the ideologies of anti-goat environmentalism proliferated in the empire. The colonial interests for commercial forestry and timber exploitation inevitably led to conflict between the colonial foresters and the goat that consummated with goat control legislation. In Cyprus, the successive British colonial foresters spoke vituperatively of goats and were highly critical of their presence in forests. In 1870, forester Gerrard de Montrichard reported that Cyprus had endured a loss of one-third of its forestry productivity as a result of the goat whose numbers exceeded that of any in the Mediterranean (Pyne 1997 133). One forester criticised the goat as "that eternal starveling whose cruel teeth and poisonous saliva was destroying Cyprus as completely as Rome's legions had done to Carthage" (Pyne 1997 136). The goat law of 1913 allowed villagers to exclude goats from communal lands and substitute them with sheep. During the 1940s, the Cyprus government reduced the number of goats by encouraging goat-herders to give up their goats in exchange for £60 smallholdings and a quarter of an acre of land for every 10 goats surrendered to the forestry department. In 1937, Greece also passed a law to remove goats from the districts in which there were most harmful within ten years (Maher 1945). In the USA, a report by forester William Zeh on the conditions of the native Indian Reservation of Navajo blamed soil erosion and land degradation on a large number of sheep and goats (McPherson 1998). Consequently, a stock reduction programme was launched which resulted in the forced sale of 148,000 goats and 50,000 sheep in 1934.

In Africa, forests were at the centre of colonial control and domination, and narratives of forestry preservation were deployed for the purposes of appropriating land resources, social control, and commodity production (Davis 2007; Ford 2008). In colonial Algeria, French concerns about forestry were linked to declensionist environmental narratives that arose in the 1850s. These narratives viewed Algerian forests as sites of memory for North Africa's Roman glory that had been ravaged by the pastoral Arab invasions and their goats and camels from the 11th century (Ford 2008). The 1827 forestry code forbade the grazing of sheep and goats in public forests and only allowed pigs. The French forestry concerns, however, were motivated by the desire to expropriate forested lands for commercial timber exploitation. Between 1890 and 1940, the extensive cork forests in Algeria had been decimated. Meanwhile, by the end of the colonial period, the nomadic pastoralist economy of Algeria had been reduced from 85% to 5%, and the number of livestock reduced by half (Davis 2004).

From the 1930s, colonial legislation had centralised land use and pushed Africans to the reserves where population pressure led to the problems of overgrazing and depletion of pastures. African goats were singled out by colonial conservation experts as "the worst offenders" (Hall 1936). The colonial officials, however, reckoned that the evils caused by the goat were largely due to the primitive nature of African goat keeping, lack of selective breeding for milk and meat, lack of grazing control, and lack of scientific ideas about soil erosion and vegetation deterioration (Maher 1945). An agricultural official in a rundown Kenyan reserve during the 1930s described goats as a "mixed curse" and a "menace" (Maher 1945). In 1942, the Soil Conservation Committee appointed to report the situation on desertification and rural water supply in Sudan noted with concern the huge number of goats in the colony (5,000,000) and pointed out that there seemed to be no means of limiting their numbers short of "forced slaughter" as they could not be "taxed out of existence" (Maher 1946). The report blamed goats for the desiccation of town outskirts and the increasing momentum in soil erosion. The report also faulted flocks of goats for causing winter dust storms. In southern Africa, ardent colonial conservationist singled out the goat as being responsible for a great deal of soil erosion (Hornby and Van Rensburg 1948; Oates 1956). Colonial conservation officials often argued that African goats were "uneconomic," and their numbers could be reduced without imperilling livelihoods and the African subsistence economy. However, goat reductions did affect African livelihoods. In 1939, the Civil Secretary in Sudan had spoken against discriminatory taxation against goats and the medical department was strongly opposed to goat control measures because they had resulted in a reduction in the supply of milk for the poor peasants (Maher 1946). Thus, goat controls in colonial Africa were part of governments' environmental narrative to marginalise the less powerful, appropriate environmental goods and racially exclude Africans from forests and fertile lands in what Michael Goldman termed "eco-governmentality" (Goldman 2001).

Navigating Colonial Boundaries: African Goats in Early Colonial Southern Rhodesia, 1892–1920

The goat has limited visibility in Southern Rhodesian agricultural archives. Unlike other livestock which are subject to much policy on commercial production, breeding, disease control, and scientific research, goats receive lukewarm attention as a commercial agricultural enterprise.[4] Indeed, white farmers did keep goats but these were relatively smaller flocks for the supply of meat rations to African labour. Africans had larger flocks that roamed freely and browsed in forests. The unfettered and highly mobile African goat stirred great worry amongst the British South African Company (BSAC) forestry officials over the destruction of timber resources. Resultantly, in 1892, the Company published forestry regulations for Mashonaland that were extrapolated from the 1888 Cape of Good Hope forestry law. These regulations gave the Conservator of forests the authority to issue licences for the exploitation of forestry products such as firewood, the grazing and pasturing of livestock, and hunting of game and wildlife (*Rhodesia Herald*, 29 October 1892). The regulations specifically forbade the grazing of goats in wooded parts of the Company's forests and in parts where young trees were growing. African goats were also viewed as unhygienic, unsanitary, and a nuisance to white residential spaces. In May 1893, the sanitary board of Victoria town allowed all other animals except goats and donkeys to graze within the perimeters of the township boundaries from sunrise to sunset (*Rhodesia Herald*, 18 March 1893). The African goat was branded as a beast of notoriety often seen ravaging yards and trampling on green gardens – a gardener's nightmare. A writer in the *Bulawayo Chronicle* fretted that the goat was a "pest of the deepest dye" and the locust was "angelic" in comparison to the goat (*Bulawayo Chronicle*, 8 August 1896). Although he compassionately noted the important role of the goat to poor families, he insisted the "wretched animals" had to be reined:

> The right to keep a few goats in the town at the time is a great boon to many poor families, but at the same time people should distinctly understand that their goats must be herded throughout the day on the commonage. Gardeners cannot forever be watching the wretched animals.

Another writer commiserated with "those working people who wisely keep a few goats to enable their children to obtain a little milk" but emphasised that it was not at all necessary for their goats to live by vandalising the few remaining shrubs and trees in their neighbours' gardens (*Bulawayo*

Chronicle, 20 June 1896). To ensure that the African goat did not stray into urban precincts, bye-laws were enacted by town councils prohibiting animals from getting into habited streets or coming into proximity with human habitations.[5] Urban sanitary boards were given powers to effect impounds of stray animals and goats which were to be auctioned if the owners failed to pay a fee. Although impounds were generally enforced and affected all livestock, African goats disproportionately suffered because of their aptitude for greater mobility crossing various spaces foraging for rubbish and titbits in human settlements. The following are some of the impound notes for African goats that were occasionally circulated in The Rhodesia Herald.

> THE following animals having been impounded, will be sold by Auction at the Pound Kraal, Sunridge, Plumtree Station, on Saturday, June 5th, 1909, at 10 o'clock in the forenoon, unless previously released:—
> Six Native Ewe Goats.
> Seven Goat Kids.
> W. D. ESTMENT,
> Pound Master.

> THE following animals having been impounded, will be sold by Auction at the Pound Kraal, Sunridge, Plumtree Station, on Saturday, September 4th, 1909, at 10 o'clock in the forenoon, unless previously released:—
> 26 Native Goats.
> 6 Native Sheep.
> W. D. ESTMENT,
> Poundmaster.

Figures 3.1a and *3.1b* Impound notes of stray African goats (The *Rhodesia Herald* 1 June and 18 August 1909)

African goats were also viewed as unhygienic and vectors for the spread of diseases. In July 1896, Mr Norton is reported to have had a small flock of goats which he milked. The sanitary board of Bulawayo condemned his milk as "inferior, if not harmful" for consumption (*Bulawayo Chronicle*, 25 July 1896). From 1905 to 1908, there was an outbreak of scab disease that affected many goats and sheep in the colony. Colonial officials were worried about the effects of the disease on the supply of sheep and goat meat to the mines for native rations. A speaker in the Legislative Assembly, Mr Peele, however, blamed the spread of the disease on "natives" tramping districts selling their goats (*Bulawayo Chronicle*, 6 November 1908). Goat meat was a cheap source of protein for "native rations" in the mines and farms, but it was not fancied by Europeans. African goats were considered scrubs with high precocity and much inclined towards promiscuous in breeding which in turn accounted for their smaller size and poor-quality meat.[6]

The numbers of African goats rose despite the restrictions on grazing and movements. The *Bulawayo Chronicle* reported that African goats were plentiful as the "natives" seldom part with them (*Bulawayo Chronicle*, 29 April 1899). The number of African-owned goats in Bulawayo rose from 68,000 in 1901 to 157,500 in 1905 (*Bulawayo Chronicle*, 3 June 1905). In Mashonaland, the figures rose from 193,012 to 223,544 during the same period. In the Victoria district, they were 180,000 goats. By 1908, the colony had a total of 593,860 African-owned goats (*Rhodesia Herald*, 11 March 1910). A large number of goats created conflicts between African goat owners and European farmers as goats trespassed into farmlands and damaged crops. For instance, in 1911, goats belonging to a Fingo native were reported to have strayed into the tobacco fields of a Greek planter in the Reserves of Induna and damaged tobacco resulting in a physical confrontation that consummated in the native being fined £2 damages.[7]

The African goat could not be confined to the geographic boundaries set up by colonial authorities to exclude and exploit Africans. The imperious goat defied the restrictive regulations, trespassed into the colonial realms of whiteness and traversed forbidden landscapes defiling their purity, sanitation, and hygiene. For this reason, the African goat was marked as an anathema and nemesis to progressive farming and land settlement for whites. It was a creature marked for slaughter to conveniently supply meat to African labourers in the mines. Meanwhile, the white settlers had to look for alternative breeds of superior goats to domesticate, and they found their answer in the Angora[8] and Boer goats.

The Angora goat had been introduced at the Cape in South Africa to produce mohair in 1838. The mohair industry had significantly grown such that by 1912, the country had a stock of 4.4 million Angora goats and produced 9.21 million kilograms of Mohair (Pringle 2011). Rhodesian settlers started importing Angora goats from South Africa as early as 1896.[9] Districts in the eastern highlands such as Melsetter, Umtali, and Inyanga were ideal settlements for rearing Angora goats. These areas had high altitude, numerous kopjes that afforded a diversified feed, good grazing, dwarfed bushes, and well-drained soils (Ewing 1905). By 1905, one white farmer, Mr Ewing, was reported to have a flock of 81 Angora goats. The Angora goat stood out as a superior specie to the African goat in a way that reinforced the conception of race and discrimination. Its physical appearance of long quality hair of a soft and silky texture covering the whole body accentuated its privileged position as not just a goat but a superior goat. The Angora goat was described in the *Rhodesia Agricultural Journal* as "nearer in nature, form and habits to the sheep" and having a remarkable degree of gentleness (Ewing 1905). When juxtaposed to the African goat that had a notorious reputation of stripping bushes bare, raiding gardens, and vandalising neat streets, the Angora goat was depicted as a benevolent browser neatly trimming bushes for the regeneration of veld and pasture improvement.

The ecological narrative obliquely peddled by colonial officials was that where African goats damaged bushes and forestry resources, the Angora goat was useful for clearing bushes of undergrowth and weeds "transforming in a remarkably short time the roughest land into beautiful pasture" (Ewing 1905). Angora goats would manure the land, increase its fertility for cultivation and regenerate the veld with new grasses. Whereas the browsing habits of the African goat had been faulted for destroying young bushes and stunting forest regeneration by the forestry ordinance, the browsing habits of the Angora goat were viewed as an asset. It solved the problem of winter feeding, and by cutting through many "native trees of no value" the goats could be kept throughout the season with no extra feeding (Ewing 1905). Thus, highly valued for their lucrative wool and mohair the Angora goats were framed in a different ecological discourse than the indigenous Mashona and Ndebele goats. Just as the development of the wool trade in medieval Europe had relegated goats to the peripheries of pastures and elevated sheep to privileged grazing lands and ecological zones the imperative for wool production in Southern Rhodesia witnessed a pastoral orientation focusing on Angora goats and Merino sheep. In 1905, Southern Rhodesian wool had made its first appearance on the London market when 7 bales of Merino sheep wool from Mr Steyn of Melsetter were disposed of at 7¼/per lb (*Rhodesia Agricultural Journal* 1905). It was emphasised thereafter that for the prosperity of the wool industry every effort and attempt was supposed to be towards preserving the purity of the breeds and excluding any chance of crossing with native sheep and goats. By 1908, mohair production was also now being done in the Gwelo district while Merino wool production was centred in the Melsetter district. Samples of Rhodesian mohair were sent to wool brokers in London for valuation (*Rhodesia Agricultural Journal* 1908). The opinion of the valuators was that the mohair was of desirable class and compared favourably with "Cape super"[10] and the class was on high demand on the London market.

Consequently, wool production came to preoccupy Rhodesian goat farmers. In 1914, the colony of Southern Rhodesia had 324,000 sheep with

Table 3.1 Statistics of goats in Southern Rhodesia, 1911–1917.

Year	Goats owned by Africans	Goats owned by Europeans	Total number of goats
1911	576,898	24,737	601,635
1914	639,000	35,000	674,790
1915	661,867	26,518	688,385
1917	735,670	30,731	766,401

80% of those owned by Africans with European farmers only owning only 67,000 sheep (Bridger 1959). The adoption of intensive farming methods such as rotational grazing for internal parasite control, and the use of breeds of sheep[11] better suited to the climatic conditions of the country witnessed a boom in the European flock from 67,000 in 1914 to 130,000 in 1957. Meanwhile African-owned sheep population fell by nearly 50% between 1914 and 1952 largely because of land centralisation laws such as the Land Apportionment Act of 1930 and the Native Land Husbandry Act of 1951. With dwindling pasturage and grazing land Africans had to choose cattle and goats over sheep, the former for their great economic value and the latter for their ability to survive harsh range conditions.[12] Table 3.1 shows that in 1911, there were 601,635 goats. Africans owned 576,898 and Europeans 24,737. By 1917, the total number of goats was 766,401 with Africans owning 720, 670 goats.

Colonial Conservation and the Goat Question in Southern Rhodesia, 1930–1968

In October 1938, an African writer in the *Bantu Mirror* Henry Ncube described the deplorable conditions for livestock in the Native Reserves wrought by colonial land centralisation and conservation schemes that had been progenies of the Land Apportionment Act. He observed, "the veld is grassless, and the river is waterless, yet the area is overstocked. The result then is that many a goat and cow die . . . the cattle are as thin as hungry rats. There are too many animals to graze on so small an area" (*Bantu Mirror*, 22 October 1938). The conditions of overgrazing and land degradation caused by overstocking were becoming endemic to most reserves during the 1930s because of population pressure on limited land resources. By the mid-1930s, the impact of overstocking was beginning to be felt in most reserves raising alarm amongst colonial officials. In 1935, Native Commissioners in newly acquired native areas were issued with powers to restrict the numbers and kinds of stock to be depastured. Donkeys and goats were the most affected by the directive as their numbers were limited to 2 and 12 for each family, respectively (National Archives of Zimbabwe NAZ S3001/1). In 1936, the agriculturalist for the Native Department expressed dismay at the levels of overstocking in the reserves which were causing soil erosion and intimated that the resolution to the ecological crisis would be a reduction in the number of African goats. The Native Department complained that goats had been permitted to increase in the Reserves and rampaging flocks were now destroying range conditions for cattle and the "natives" had to make a choice between his goats and cattle (NAZ S3001/1). In light of the important economic value of cattle to Africans, the Department officials proposed the limitation of the numbers of goats through several interventions. These included enforced castration of all-male goats to a designated limit, killing all male goats but one to a designated ewe, killing all new-born twin goats, limiting the number of goats per owner, taxing all goats over a designated number, encouraging sheep rather than goats in the reserves by not taxing sheep and following up the American example in the Indian Navajo Reserves by buying up goats and slaughtering them. The officials reckoned that the enforced castration of all-male goats and the other measures would not only reduce numbers but improve quality while sheep would supply better meat and were less destructive to range conditions. The Chief Native Commissioner concurred and pointed out that the increase in donkeys and goats during the previous ten years had been cause for alarm. However, he proposed a graduated and peaceful resolution to the goat question. This included allowing a limited number of goats and donkeys into the Native Reserves, and if a tenant wished to keep more than the fixed number, a grazing fee of 6/ per head per annum was to be imposed for goats.

The Chief Native Commissioner reiterated that goats and donkeys were the most destructive animals both to grazing and forestry. He pessimistically prognosticated that if the rates of increase in African goat populations continued there was a grave danger of the western portion of the colony being reduced to "desert conditions" just like the Sinai Peninsula and the Arabian desert that were ravaged by goats for centuries (NAZ S3001/1). However, he was little enthusiastic on the complete elimination of goats in the Reserves as a few were necessary to protect sheep in areas that had wild carnivora." He recommended that after a period of 50 years only three goats were to be allowed to each flock of sheep in the Reserves except in tsetse fly-infested areas. This was to give natives time to dispose of the surplus but in the meantime, a grazing fee of 1 penny per head for all goats in excess of three was to be charged after two years' notice. Goats were also to be excluded or limited in deeds of sale to natives. The Native Department argued that goat taxation would raise money for the development and conservation of natural resources in the Reserves through the newly established Native Councils (NAZ S3001/1).

The logic that informed colonial officials' policy on the goat question in Southern Rhodesia was that goats were more destructive to the grass, trees, and shrubs and they were multiplying faster and were found in larger numbers than cattle in most reserves.[13] Furthermore, goats were perceived as being of low economic value to the Africans as few were consumed and relatively few were sold. In the mind of colonial officials, the African goat was an "uneconomic goat," very distinct from the "economic goat" found in the European farms that produced mohair for export and supplied commercial meat. However, goats had a bigger intrinsic value indispensable to the domestic economy and social life of Africans. Leakey in his study of the role of goats in the social life of the Kikuyu pointed out that colonial arguments on the economic dispensability of the goat largely ignored the "sociological value" of the stock to tribal life (Leakey 1934). Reductions on goat herds were inclined to cause a massive disruption to tribal life as these animals carried social capital through their use in marriage insurance, payments, sacrifices, festivals, purification ceremonies, and spiritual cleansings. In Southern Rhodesia, goats were also referred to as "the tribal refrigerator" as they were usually slaughtered to provide meat for special occasions (Cross 1974). Many of the goats were owned by women and children and were important assets in bestowing property rights to subaltern groups in African societies. Consequently, women and children were largely opposed to the sale and reduction in the number of goats. When trading stations were set up in the reserves to coerce Africans to sell their goats very few were forthcoming and the situation was described by the Native Commissioner of Plumtree as more complex than it appeared since "nearly every married woman and not a few children are now stock owners and resent any interference by the elders in their property rights" (NAZ S3001/1). In Bulawayo, a trading station for goats that had been set up in 1938 only managed to purchase 130 African goats out of 30,000. These low trading figures were attributed to difficulties that the "native" had to contend with in disposing goats – their women! The Superintendent for Natives reported that women who had been deprived of their "useless stock" (meaning goats) would disturb the peace of the home and local community for days and their influence was very extensive (NAZ S3001/1).

However, although goats were included in the destocking programme in 1938 on the grounds of the damage done by them to grasslands the Native Trade and Industrial Commission of 1944 known as the Godlonton Report noted the valuable potential supplement provided by their milk and meat and recommended that limitations on goats were to be imposed sparingly (The Native Trade and Industrial Commission of Southern Rhodesia 1944). Native Land Husbandry Act (NLHA 1951) however, significantly affected the population of African goats. The Act provided for the reduction of livestock numbers in Tribal Trust Lands nearer to the assessed carrying capacity under approved livestock units.[14] *The Report of the Advisory Committee of the Development of the Economic Resources of Southern Rhodesia with Particular Reference to the Role of African Agriculture* (1962) pointed out that Africans preferred to retain their cattle and sell their goats despite the fact that goats provided greater security and a much greater scope for turnover (Phillips et al. 1962: 177). The report observed the important role of goats in African areas and advised that Africans in Tribal Trust Lands and

African Goats, the State and Conservation   49

arid areas were supposed to be actively encouraged to keep more goats as these animals were resilient browsers. However, the report added a caveat that unless their numbers were strictly controlled, goats would "do more harm than good." The effect of the NLHA on African goats was deleterious. While there were over one million sheep and goats at the end of the 1930s, there were only 131,000 sheep and 380,000 goats in 1960 (Phillips et al. 1962: 177).

In 1962, the destocking programme in the native reserves came to an end when the Native Land Husbandry Act was repealed. The effect was a huge surge in the number of African goats. By 1974, estimates of the number of goats in African areas ranged from 1.6 million to 3.5 million, while the goat herd on European-owned farms was 35,000 (Cross 1974). African goats came to attract significant commercial interests from the state meat-processing company Cold Storage Commission (CSC) and private enterprises who began to buy, process, and sell goat meat.[15] At the same time, the slaughter of goats for consumption in the tribal areas increased significantly because of population growth. The tribal slaughter of goats doubled from 1968 and was estimated to exceed 600,000 heads a year, or 40% of the official goat output (Cross 1974). The rise in tribal consumption of goat meat was now seen as a threat to prospects for commercial slaughter. The Agricultural Marketing Authority of Southern Rhodesia noted that the market for goat meat was largely found in communities with "unsophisticated consumption requirements"[16] such as the African, Asian and Coloured communities in Southern Rhodesia and the Coloured and Indian communities of Natal and the Cape in South Africa (Cross 1974). The investigation of the goat industry by the Agricultural Marketing Authority of Southern Rhodesia in 1973 concluded that local and external demand for good meat at satisfactory prices in relation to prevailing producer prices was in excess of the capacity of Southern Rhodesia's supply. The investigation report noted that production potential was being curbed by high levels of subsistence consumption and traditional attitudes towards livestock that were inhibiting sales. The investigation observed that the high demand for home consumption of goat meat constrained the sale of goats to commercial outlets despite considerable investment having already been made in facilities for the purchase and processing of goat meat. The Agricultural Marketing authority concluded that population growth in the tribal areas was threatening the viability of commercial goat production and a decline in human population in these areas due to industrialisation and urban growth was the only prospect that could increase the sale of goats to commercial processers. The report gloomily forecasted that "if the present trend of population growth persists, it is likely that the demand for goats for home consumption will not permit the sale of goats on a significant scale to commercial outlets" (Cross 1974). Table 3.2 below illustrates the figures of the commercial slaughter of African goats between 1968 and 1973.

50 Elijah Doro

Table 3.2 Commercial slaughter of African goats in Southern Rhodesia, 1968–1973.

Year	Number of Commercial Goat Slaughter
1968–69	34,982
1969–70	48,169
1970–71	38,988
1971–72	48,988
1972–73	65,777

Beyond Marginality: Goats, Pasture Management, and New Environmental Landscapes in Southern Rhodesia, 1950–1970

Between 1950 and 1970, several developments in pasture research and veld management in Southern Rhodesia and most parts of southern Africa evoked a reappraisal of the role of goats in grazing and ranching ecosystems. In 1942, Dr Pole Evans, a South African pasture researcher of remarkable merit, had been invited by the Rhodesian government to report on grazing conditions in the country and to make recommendations. His report had stressed the damage done to the countries veld and implored on the need for grass cover and recommended a comprehensive programme of pasture research. Subsequently, two pasture research stations were opened at Matopos and Marandelas in 1945. By the early 1950s, much of the veld work was being revised as research was showing that the critical essential for the improvement of sand veld grazing was the reduction in the density of trees and bush (West 1956). The problem of "bush encroachment"[17] now threatened pastures and grazing lands in Southern Rhodesia. Various methods were employed to deal with bush encroachment by cattle farmers and these included the mechanical and manual removal of bushes, fire control through the burning of the bushes, and chemical elimination using arsenic poison as an arboricide. However, manual control was labour intensive and costly, fire control was wasteful and the coppices of certain species such as Musasa (*Brachystegia Spiciformis*) and Munhondo (*Julbernalia Globiflora*) could not be controlled by burning (Strang 1973). The use of arsenic (Arsenite of Soda and Arsenic Pentoxide) as an arboricide was highly risky as it was toxic to cattle and required extreme care. Research was also showing that although cheaper than mechanical control, arboricides were still costly and of a doubtful ecological validity (Little and Ivens 1965). From the early 1960s, goats were integrated by cattle farmers into veld management as a biological means of controlling bush encroachment in grazing lands. Pasture research stations such as Matopos and Makhaholi had shown that mixing herds of goats and cattle in open pastures improved grass cover. Southern Rhodesian pasture management experts such as O. West, T.C.D. Kennan, and R.R. Staples concurred on the ecological utility of controlled use of goats to eliminate coppices (Kennan et al. 1955). They advocated careful management to ensure that the goats would not damage grass through excess trampling. Above their preference for browsing to grazing, goats were preferred because they had the extended advantage unlike cattle of being able to digest the seeds of thorn trees and limiting the further spread of thorn-bush trees (*Acacia*). Also, beyond the ecological value of integrating goats in pasture research, there was the economic benefit as goats were a useful source of protein and could be used for "native" rations as it was now shown that goats yielded more meat than cattle in terms of pound per acre (*Rhodesia Agricultural Journal* 1973).

Many opinions amongst conservationists in Southern Rhodesia now shifted from open hostility against goats towards a controlled management of goat populations and distributions as a way of limiting the deterioration of pastures and natural ground cover. A pilot project on integrating goats to stop the spread of Acacia bushes had been started at Matopos research station in 1950 (Oates 1956). After four years the results showed that grass cover within the paddock improved a great deal and areas that previously had no annual grass cover were now much smaller in size and invaded by perennial grasses. The experiment results also revealed that goats could eliminate certain trees and bushes without harm to the grasses. Other pasture research results showed that goats did not directly compete with cattle for grazing and the increase in mass showed that performance per animal was slightly more favourable when goats grazed together with cattle (*Rhodesia Agricultural Journal* 1973). An earlier study by Staples et al. (1942) in Tanganyika to compare the effects of introducing goats into mixed-grass bush pastures together with cattle had also revealed that goats' preference to browsing kept them from causing erosion where there was sufficient grass cover but in fact resulted in pastures constituting "good grazing for any kind of local stock."

Conclusion

Despite their marginality as subaltern livestock, African goats had the ubiquity and mobility that enabled them to trespass into forbidden spaces and boundaries imposed by the colonial state-restructuring physical geographies and transforming rural livelihoods. The African goat contested colonial conceptions of vernacular landscapes and African livestock regimes in ways that illuminated dispossession, marginalisation, social disruption, and environmental racism in much more compelling narratives because of their proximity to the weaker and marginalised social groups. Therefore, the goat affords a panoptic view of the bigotry and racial scapegoating that informed colonial conservation ideologies and the framing of environmental discourses in colonial Africa. Unfortunately, these hegemonic colonial racist environmental and conservation constructions have remained preponderant in the framing of post-colonial livestock and environmental discourses in Africa where goats and indigenous livestock regimes remain relegated to the peripheries of conservation, sustainable development, and climate action. Consequently, there have been very few successful goat breeding programmes in post-colonial Africa as state agricultural planners and researchers have failed to transcend the historical neglect and bigotry of colonial officials (Peacock 2005). Agricultural programmes in Africa have not been robust in improving breeds and harnessing goats into long-term strategies of growing rural economies and strengthening resilience to climate change amongst historically marginalised communities that stand to benefit from more pro-active integration of goats into climate policy action. Resultantly, there has been little initiative on commercial goat farming in Africa and goats have remained at best a subsistence household initiative. In post-colonial Zimbabwe, for instance, during the Rural Land Resettlement Program of the 1980s and 1990s goats were prohibited in the new African rural resettlement areas and were considered a nuisance to planned land settlement. Their spaces remained constricted and ghettoised to the colonially constructed Communal Lands where there were scarce land and environmental resources. Resultantly, goat populations in most of the country's agricultural zones remained stunted and agrarian studies unsurprisingly point towards higher cattle holdings and contend that goats are "not common on any site" and "relatively few are kept as a household flock" (Shonhe et al. 2020). In the end, therefore, there is a need to re-imagine goats within new post-colonial environmental landscapes and within a climate-justice oriented framework that views them beyond the colonial bigotry and racial scapegoating. This is because the potential for goats to construct luxuriant ecologies and sustain livelihoods is much conspicuous despite their limited temporal visibility. The indigenous goat can be harnessed as an important component of sustainable development, environmental management, and poverty eradication in Africa, particularly in areas with adverse environmental conditions and amongst historically marginalised climate and food insecure groups. Goats can also be integrated into urban ecologies and landscapes through sustainable waste management and peri-urban goat farms that can create recreational spaces and centres for urban tourism (Richardson and Whitney 1995). Ecologists have also argued that goats can be useful for controlling fire outbreaks caused by global warming and climate change and have attributed the prevalence of contemporary forest fires to the declining pastoral economies in forested regions (Siddle 2009).

Notes

1 Classical writers such as Virgil referred to the dangers of overgrazing by goats and the need to keep them away from tender pastures.
2 The Island of St Helena has been touted as a classic example of the ecological catastrophe resulting from the depredation of goats. Originally forested the Island had been inhabited by the British who introduced goats in the 17th century for supplies of meat. Heavy felling of trees for wood fuel depleted forestry

resources and regeneration of the forest was stunted by the browsing of goats. By 1810, St Helena had been reduced to a wasteland with extensive erosion. Scientific studies on the effects of goats on island ecosystems point to alteration of vegetation and habitat destruction resulting in extinction of endemic bird and snake species.

3 Graeco-Judean Christian culture portrays the goat as a symbol of the devil and representing sinners. In African culture goats are often used as sin offerings. In Zimbabwe for instance the month of November is tabooed and named *mbudzi* which literally translate to goat(s). During this month, it is still considered sacred to conduct any traditional rituals such as marriage or paying homage to ancestors.

4 *The Rhodesia Agricultural Journal* which was the technical bulletin for white commercial agriculture from as early as 1903 discusses various aspects of animal husbandry and has extensive articles on cattle production, pig farming, poultry and sheep keeping but rarely engages with the subject of commercial goat keeping. Goat keeping as a subject only features once in all the entries in the journal between 1903 and 1955.

5 Notices in *Rhodesia Herald* from 1892–1896 contain various bye laws by towns to contain movement of domestic animals into inhabited areas.

6 These are the opinions expressed by Mr W.E. Scott, Native Commissioner for Lomagundi in the *Rhodesia Agricultural Journal* in 1904.

7 Laws against stray goats also existed in South Africa. A civil case is recorded in the Rhodesia Herald where a man from Potchefstroom had his goats impounded after they broke into a government reserve and destroyed a number of Cypress trees. See 'Goats in Court', *Rhodesia Herald*, 18 June 1909.

8 Angora goats are descendants of Persian wild goats (*Hircus aegagrus*) and originate from the Turkish Province of Ankara. They are prized for their long and silky hair which is commonly referred to as mohair. Mohair is used in the industrial manufacturing of clothes, drapers, furniture, upholstery and carpets.

9 *Bulawayo Chronicle* of 8 February 1896 reported that Angora goats were being dispatched from the Cape to Rhodesia to provide a nucleus for a profitable mohair industry.

10 Phrase used to describe the quality mohair produced by goat farmers at the Cape in South Africa.

11 Ewes and rams were imported from South Africa. In 1936, £1,000 was provided by white farmers to that endeavour.

12 Colonial officials expressed dismay that Africans could chose to keep cattle and goats rather than sheep since sheep according to them were better suited to production within small holdings and to producers with little capital.

13 The Report of the Commission of Enquiry in Certain Sales of Native Cattle in Areas Occupied by Natives (939) whose terms of reference were to report on actions being taken by the government to reduce the number of cattle in areas of the colony set aside for use and occupation by natives and methods carried out by government to reduce the number of native cattle noted the huge number of goats in some districts. In Victoria district for instance, the report noted that there were 37,300 goats against 33,595 cattle.

14 A livestock unit comprised five stock (goats or sheep) or one cattle unit.

15 The *Report of the Advisory Committee of the Development of the Economic Resources of Southern Rhodesia with Particular Reference to the Role of African Agriculture* noted that substantial numbers of sheep and goats were sold and consumed in towns and African areas.

16 Goat meat was not popular amongst Europeans.

17 'Bush encroachment' refers to the invasion of a grass dominated community by woody species. The grasses are suppressed and their foliage production is reduced. This usually depleted pastures and threatened the cattle industry.

References

Primary Sources

Bantu Mirror. 22 October 1938.
Bulawayo Chronicle. 1896, 1899, 1905, 1908.
National Archives of Zimbabwe, S3001/1, Agriculturalist Native Department to Chief Native Commissioner, 15 December 1936.
National Archives of Zimbabwe, S3001/1, Cattle Correspondences, Chief Native Commissioner to Prime minister, 19 December 1936.
National Archives of Zimbabwe, S3001/1, Cattle Correspondences, NAZ, S3001/1, Native Commissioner Plumtree to Superintendent of Natives, 2 August 1938.
National Archives of Zimbabwe, S3001/1, Native Commissioner Plumtree to Superintendent of Natives, 30 August 1938.
National Archives of Zimbabwe, S3001/1, Superintendent of Natives Bulawayo to Chief Native Commissioner, 1 September 1938.
Rhodesia Herald. 1892, 1893, 1909, 1910.

Secondary Sources

1904. Precocity in sheep and goats. *Rhodesia Agricultural Journal* 1:177.
1908. Merino wool and mohair. *Rhodesia Agricultural Journal* 3:264–265.
Badenhorst, S. 2002. The ethnography, archaeology, rock art and history of goats (*Capra Hircus*) in southern Africa: An overview. *Anthropology Southern Africa* 25:96–103.
Badenhorst, S. 2006. Goats (*Capra hircus*), the khoekhoen and pastoralism: Current evidence from Southern Africa. *The African Archaeological Review* 23:45–53.
Bridger, G. A. 1959. Sheep production in Southern Rhodesia. *Rhodesia Agricultural Journal* 56:211–217.
Brink, J. S. and S. Holt. 1992. A small goat, *Capra hircus*, from a late iron age site in the eastern Orange Free State. *South African Field Archaeology* 1:88–89.
Cross, E. G. 1974. Goat marketing in Rhodesia. *Rhodesia Agricultural Journal* 71:159–160.
Davis, K. 2004. Eco-Governance in French Algeria: Environmental history, policy and colonial administration. *Journal of the Western Society of French History* 32:328–345.
Davis, K. 2007. *Resurrecting the Granary of Rome: Environmental History and French Colonial Expansion in North* Africa. Athens, OH: Ohio University Press.
Digests. 1973. *Rhodesia Agricultural Journal* 70:92.
Dohner, J. 2001. *The Encyclopaedia of Historic and Endangered Livestock and Poultry Breeds*. New Haven: Yale University Press.
Dwyer, C. 2017. The behaviour of sheep and goats. In *The Ethology of Domestic Animals: An Introductory Text*, ed. P. Jensen. Boston: CAB.
Ewing, W. L. 1905. The Angora goat. *Rhodesia Agricultural Journal* 3:129–130.
Ford, C. 2008. Reforestation, landscape conservation, and the anxieties of empire in French colonial Algeria. *The American Historical Review* 113:341–362.
Goldman, M. 2001. Constructing an environmental state: Eco-governmentality and other practices of a 'green' world bank. *Social Problems* 48:499–523.
Hall, D. 1936. *The Improvement of Native Agriculture in Relation to Population and Public Health*. London: Oxford University Press.
Hornby, H. E. and H. J. Van Rensburg. 1948. The place of goats in Tanganyika farming systems. *The East African Agricultural Journal* 14:94–98.

Kennan, T. C. D., R. R. Staples and O. West. 1955. Veld management in southern Rhodesia. *Rhodesia Agricultural Journal* 52:4–21.

Leakey, L. S. B. 1934. Some problems arising from the part played by goats and sheep in the social life of the Kikuyu. *Journal of the Royal African Society* 30 (130):70–79.

Little, E. C. S. and G. W. Ivens. 1965. The control of brush by herbicides in tropical and sub-tropical grasslands. *Pest Articles and News Summaries* 11 (3):245–262.

Maher, C. 1945. The goat: Friend or foe. *The East African Agricultural Journal* 11:115–121.

Maher, C. 1946. Goats fire and blowing sand. *The East African Agricultural Journal* 11:173–180.

McPherson, R. 1998. Navajo livestock reduction in southeastern Utah, 1933–46: History repeats itself'. *American Indian Quarterly* 22:1–18.

Mwatwara, S. 2014. A history of state veterinary services and African livestock regimes in colonial Zimbabwe, c.1896–1980. PhD diss., Stellenbosch University.

Oates, A. V. 1956. Goats as a possible weapon in the control of thorn bush. *Rhodesia Agricultural Journal* 53:68–85.

Peacock, C. 2005. *Goats: Unlocking Their Potential to African Farmers*. Paper presented at the seventh Conference of Ministers responsible for animal resources, Kigali.

Phillips, J. et al. 1962. *Report of the Advisory Committee of the Development of the Economic Resources of Southern Rhodesia with Particular Reference to the Role of African Agriculture*. Salisbury: Mardon Rhodesian Printers.

Pringle, T. 2011. A history of the South African mohair industry 1838–1971. *South African Journal of Economic History* 4:55–77.

Pyne, S. 1997. *Vestal Fire: An Environmental History Told Through Fire of Europe and Europe's Encounter with the World*. Seattle: University of Washington Press.

Richardson, G. M. and J. B. R. Whitney. 1995. Goats and garbage in Khartoum, Sudan: A study of the urban ecology of animal keeping. *Human Ecology* 23:455–475.

Robinson, K. R. 1986. A note on the occurrence of goats and sheep in the rock art of north-eastern Zimbabwe. *The South African Archaeological Bulletin* 41:32–34.

Samasuwo, N. 2000. There is something about cattle: Towards an economic history of the beef industry in colonial Zimbabwe, with special reference to the role of the state, 1939–1980. PhD diss., University of Cape Town.

Segui, E. 1946. La Guerre aux Chèvres sous l'Ancien Régime. *Cahiers d'histoire et d'archaéologue* 1:11–21.

Shonhe, T., I. Scoones and F. Murimbarimba. 2020. Medium scale commercial agriculture in Zimbabwe: The experience of A2 farms. *Journal of Modern African Studies* 58:601–626.

Siddle, D. 2009. Goats, marginality and the dangerous other. *Environment and History* 15:521–536.

Staples, R., H. E. Hornby and R. M. Hornby. 1942. A study of the comparative effects of goats and cattle on a mixed grass-bush pasture. *The East African Agricultural Journal* 62:62.

Strang, R. M. 1973. Bush encroachment and veld management in south-central Africa: The need for a reappraisal. *Biological Conservation* 5:96–104.

Virgil. 1969. *Georgics*. London: Folio Society.

West, O. 1956. Pasture improvement in the higher rainfall regions of southern Rhodesia. *Rhodesia Agricultural Journal* 53:439–451.

4 The Politics of Exclusion and Violence in Protected Areas

Tafadzwa Mushonga

1. Introduction

Nature reserves are defended as territories for biodiversity conservation (Terborgh 2004; Dudley and Stolton 2010). Time, events, and processes have, however, shown that resource-rich enclosures support a particular capitalist network of powerful public and private actors exploiting resources in the name of conservation while excluding those whose extraction interests are perceived as a threat (Brockington et al. 2008). Aspirations to profiteer from resources and the need to exclude "others," have necessitated the intensification of violent exclusionary measures in conservation areas with valuable resources (Bocarejo and Ojeda 2016; Kelly and Ybarra 2016; Massé and Lunstrum 2016). Such violent practices have ignited a protracted human welfare debate in addition to concerns over the future of conservation (Brockington et al. 2006; West et al. 2006; Wilkie et al. 2006; Duffy et al. 2019).

Against these scholarly developments, this chapter examines the politics of exclusion and violence in resource extraction. Using the case of Sikumi Forest Reserve in north-western Zimbabwe, it pays attention to the interests of the array of actors in the resource-rich protected area, and how such interests become the basis of private and public actor coalitions of resource extraction, and for violent othering. It discusses the effect of violent exclusionary approaches, and why such a *modus operandi* could be disadvantageous for present and future conservation efforts in resource-rich frontiers.

The analysis is set within a broad political ecology theoretical framework and supported by evidence drawn from interviews as well as lived experiences in Sikumi Forest between April 2016 and November 2017. It is additionally supported by secondary data and several years of working experience with the Forestry Commission (FC), the state authority responsible for the management of all state forests in Zimbabwe. I begin by engaging with an existing body of literature on territorialisation, securitisation, and militarisation – processes that theoretically explain how politics of exclusion and violence are enabled in resource-rich nature reserves. Next, I provide a brief resource context of Sikumi before turning to the extractive interest of

DOI: 10.4324/9781003287933-4

The Politics of Exclusion and Violence in Protected Areas 57

different actors in the reserve. From converging interests, I then touch on the areas of divergence to specifically demonstrate how joint violent exclusionary efforts are orchestrated and executed by dominant alliances. I argue that violent exclusionary practice is not entirely a conservation issue; it is also a human rights issue with counterproductive effects on biodiversity conservation.

2. Linking Resource Extraction, Exclusion, and Violence to Territorialisation, Securitisation, and Militarisation

Understanding how the power of inclusion and exclusion is generated, and how violence is perpetrated in frontiers earmarked for conservation and resource utilisation begins with understanding the process of territorialisation. Vandergeest and Peluso (1995) show how territorialisation is not simply a scientific, rather a political process of creating territories involving the demarcation of physical boundaries governed and controlled by a set of institutions, which are reinforced by law enforcement.

In respect of areas such as Sikumi Forest, territorialisation was also a process of green colonialism (Kwashirai 2009). Kwashirai particularly shows how green colonialism was enabled by racial categorisation and segregation in resource-rich geographies. The introduction of forest and wildlife law in Zimbabwe and most parts of southern Africa, for example, further institutionalised segregatory resource access allowing predominantly white-owned private timber and hunting concessions to conduct harvesting activities in forest areas, while criminalising access activities of black locals. Commenting on the role of conservation law, Peluso and Vandergeest (2001) argue that such legal framing of forests is the structural basis for excluding or including people in resource frontiers designated as state property, and, of course, a tool for state power. In their very character, territories like Sikumi Forest take the form of what Peluso and Vandergeest frame as political forests.

Literature shows that territorialisation has not been without agency. Over time, people excluded from resource frontiers have found ways of covertly or overtly transgressing territorial borders and resisting conservation law (see, for example, Guha 2000; Given 2002; Spierenburg et al. 2006; Holmes 2014; Matose 2014). The rise of illegal armed ivory and rhino poaching in the past two decades has been, perhaps, the most concerning form of resistance presented before conservationists. In Zimbabwean and South African national parks, for example, transgressors are reported to be armed Mozambicans and Zambians threatening national heritage, territorial sovereignty, and security (Duffy 2010). Although with little evidence, a narrative that ivory poaching funds terror groups such as Boko Haram and Al Shabab have further generated panic around regional and global security (Duffy 2016; Duffy et al. 2019). Elsewhere, it has been argued that the poaching crisis also, in fact, threatens the dreams of white supremacy in

Africa (Büscher and Ramutsindela 2016). Together, these dynamics further complicate the politics around frontiers such as Sikumi Forest.

The poaching crisis has, in many ways, particularly turned nature reserves into sites of security threat (Lunstrum and Ybarra 2018). Massé and Lunstrum (2016: 227) observe that this has led to "extraordinary security measures." As commercial poachers advance in their tact, law enforcement in these areas is no longer simply territorial but a security measure dispensed by military and paramilitary personnel. This security personnel permeates military culture and technology into conservation enforcement, shaping what has now been framed as the militarisation of conservation (Duffy 2017), green militarisation (Lunstrum 2014), or green violence (Büscher and Ramutsindela 2016). The result has been the production of conservation territories characterised by deadly violence (Neumann 2004).

While the need to territorialise, securitise, and militarise protected areas has been largely defended on the basis of biodiversity conservation and national sovereignty, it is increasingly incentivised by commercial gains (Ojeda 2012; Duffy 2016; Massé and Lunstrum 2016). Massé and Lunstrum (2016: 227) have developed the concept of accumulation by securitisation to elaborate, "the ways in which capital accumulation, often tied to land and resource enclosure, is enabled by practices and logics of security." Accumulation by securitisation brings together conceptual aspects of territorialisation, securitisation, and militarisation at the same time advancing critical literature on neoliberal conservation, green grabbing, green security, and accumulation by dispossession (Benjaminsen and Bryceson 2012; Fairhead et al. 2012; Büscher et al. 2014; Kelly and Ybarra 2016). This literature shows how resource-rich areas are now being territorialised with specific interest for capitalist investment. Thus, exclusion in such areas is now predominantly an economically motivated process.

What is also explicitly demonstrated in this literature is that these processes are embedded in logics of violence reminiscent of green colonisation. Brockington and Igoe (2006), for example, give an account of how the process of territorialisation involves the violent expulsion of people from protected areas – a process that has produced several conservation refugees (Dowie 2005; Lewis 2010). The art of mapping and institutionalisation of regulations exhibited in territorialisation, is in itself a systematic means of control manifesting as structural violence (Galtung 1990), while enforcement using (para)military personnel, tactics, and technologies is a practice founded on ideologies of direct violence (Feld 1977; Bernazzoli and Flint 2010). Sikumi Forest shows how these conceptual and political processes shape the politics of exclusion and violence in Zimbabwe's protected areas. The contributions gained from Sikumi Forest complement the growing critical scholarship engaging in discussions on violence and its deleterious consequences on conservation efforts. Before turning to present resource extraction politics, the author briefly situates the resource context justifying the territorialisation, securitisation, and militarisation of Sikumi Forest.

3. Resources in Sikumi Forest

Sikumi Forest is endowed with valuable resources that fall under three main categories: commercial hardwood timber, wildlife, and non-timber forest products (NTFPs). Abundance of commercial hardwood timber species is one of the chief reasons Sikumi Forest was territorialised and gazetted as a protected forest. The enclosure is located on a unique geological formation called the Kalahari sand dunes. On these dunes grows *Baikiaea plurijuga* (Teak), *Pterocarpus angolensis* (Mukwa), *Guibourtia coleosperma* (Mchibi), and *Afzelia quanzensis* (Lucky bean tree) timber species, which are of high commercial value (Bradley and McNamara 1993; Mudekwe 2007). During the colonial reservation era, these species were important for supporting rail construction and mining industries in the then company state, British South Africa Company (BSAC), but are now predominantly used in the furniture industry (Matose 1997). In association with these top commercial timber species are several other tree species adding diversity and, thus, increasing the value of Sikumi Forest (see Bradley and McNamara 1993).

What also makes Sikumi valuable is that it sits on the northeastern border of Hwange national park (Hwange), the second-largest national park in southern Africa after Kruger National Park in South Africa. Hwange and Sikumi Forest are separated by a permeable borderline, which allows wildlife to move between the two nature reserves. Thus, in addition to timber species, the forest is also rich in wildlife populations such as the famed big five as well as other large and small wildlife species. While BSAC's economic returns were from timber extraction and wildlife hunting (Kwashirai 2009), today's dominant economic activity in Sikumi Forest focuses on predominantly wildlife management and eco-tourism. To boost ecotourism, timber exploitation has been abandoned in the reserve. Lastly, the biodiversity in Sikumi Forest provides a wide range of NTFPs. The diversity of vegetation and animal species is a source of bushmeat, indigenous fruits, medicines, edible worms, and honey, while the forest undergrowth provides resources such as mushrooms, edible roots, broom grass, and grazing for livestock. Around these resources, converge a range of actors and interests. I will trace the politics of exclusion shortly. For now, here is a description of the actors and how their interests converge in Sikumi Forest.

4. Converging Interests

There are five categories of actors interested in Sikumi Forest's resources, the state represented by its agencies, the FC and National Parks and Wildlife Management Authority (ZimParks), private safari operators, international, regional, and national NGOs, commercial poachers, and local people (Figure 4.1). Each of these actors has specific interest in Sikumi's resources.

60 *Tafadzwa Mushonga*

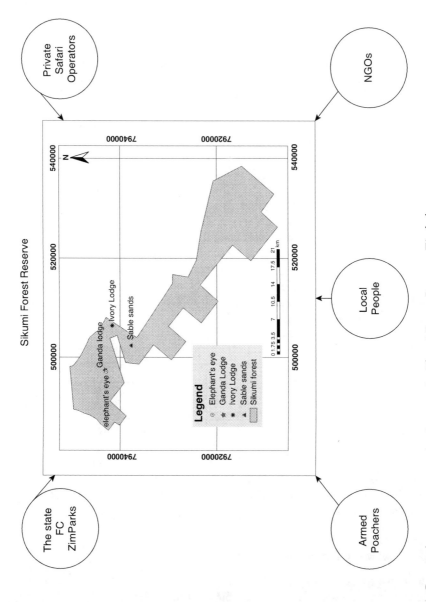

Figure 4.1 Converging actors, converging interests in Sikumi Forest Reserve, Zimbabwe.

The State

According to the official state mandate, the primary interests of state authorities in Sikumi Forest are conservation, research, and development. The FC supports its interest in conservation by arguing that forests like Sikumi are important for watershed management (Forestry Commission 2013). According to the forest authority, the Kalahari sands, on which the forest is located, are unstable soils that need to be protected against erosion to avoid the siltation of rivers that flow into Lake Kariba, the largest man-made lake and the major source of hydro-electric power in Zimbabwe. For ZimParks, Sikumi Forest is an important buffer zone, extending wildlife habitat and providing wildlife corridors for migratory animals. ZimParks, therefore, takes keen interest in the conservation of Sikumi Forest as part of its broader national wildlife management mandate.[1] In addition to being a buffer zone to Hwange, Sikumi Forest is contiguous to local communities, who, as I shall elaborate in the next few sections, are perceived as a threat to conservation. Conservation interests by the FC and ZimParks are, therefore, predominantly pitched around mitigating biodiversity degradation and extinction, which they argue to be a result of human activities.

State agents are, however, not only interested in conservation or related research and development. Their interests for conservation have since colonisation been systematically tied to resource utilisation and capital accumulation. This is demonstrated in the FC's mission statement, which is "to regulate conserve and enhance capacity in sustainable management and utilisation of forest resources."[2] Likewise, ZimParks is conserving wildlife for "sustainable utilisation of natural resources."[3] While wealth accumulation during colonial exploitation of resources was in relation to Rhode's ambitious territorialisation of all of Africa (Laterza and Sharp 2017), present-day state interest in wealth accumulation has become a survival strategy. Since independence in 1980, Zimbabwe has undergone, and still experiences, extreme economic challenges characterised by several economic adjustment policies, and reduced funding for conservation state institutions (Mushonga 2018). From as early as the 1990s, state institutions have consequently been coerced to turn to resources they manage for their survival (Matose 1997). The FC was completely weaned from state funding in January 2016. ZimParks had been weaned years earlier, thereby increasing economic dependency on resources. There is now a focus on increasing wildlife production for ecotourism, which is not only an alternative but also a more lucrative strategy for revenue generation. The FC owns four photographic safari lodges in the reserve, while eco-tourism activities by ZimParks benefit from the extension of wildlife habitat provided by Sikumi Forest.

Private Safari Operators

Private safari operators are in Sikumi Forest for capital investment in tourism. They do have an interest in conservation, but this interest is directly linked to commercial gains. For example, Elephant Eye Safari Lodge donated boots for paramilitary forests guards. It also pledged to donate a poacher's bonus of about US$ 500 and rounds of ammunition in support of anti-poaching in Sikumi Forest.[4] For private concessions like Elephant Eye, anti-poaching is important for wildlife conservation and security for their business. Anti-poaching activities serve as a guarantee to tourists that they will see animals as advertised, consequently resulting in good reviews on travel and restaurant websites such as the American-based TripAdvisor. Furthermore, private safari operators operate on a lease basis. So, in addition to pleasing holiday clients, participating in conservation initiatives is a strategic move for maintaining good relations with state authorities. Good relations secure the extension of operating licences, which is key for their business.

Non-governmental Organisations

Sikumi Forest connects wildlife from Hwange to several habitats within and outside Zimbabwe. This function has attracted the interest of the Peace Parks Foundation (PPF) – an international organisation known for running neo-liberal transboundary conservation projects in developing countries (Büscher and Ramutsindela 2016). The Peace Parks Foundation proposes to develop a boundless southern Africa with key interest in regional wealth accumulation, development through ecotourism, regional security, and peace (Andersson et al. 2013; Büscher and Ramutsindela 2016). One of its key projects in the region is the Kavango Zambezi Transfrontier Conservation Area project (KAZA-TFCA), a project that also encapsulates the Southern African Development Community's vision of regional integration through sustainable natural resources management and eco-tourism (Kavango-Zambezi Organisation 2014), and one whose frontiers include Sikumi. Thus, the interest of transboundary projects in Sikumi Forest is in conservation as much as it is in the commodification of nature and wealth accumulation.

The forest reserve also forms the framework for WWF-Zimbabwe's Hwange Sanyati Biological Corridor Project (HSBCP) launched to improve biodiversity conservation in the greater Hwange territory. The WWF is seen, here, aligned to state conservation goals, contradicting the discourse that Non-governmental organisation (NGOs) are apolitical organisations (see Fisher 1997; Duffy 2010). A hidden interest is, however, that for WWF, frontiers like Sikumi Forest provide opportunities to obtain financial resources from International Financial Institutions such as the World Bank. In the case of the HSBCP, the inclusion of Sikumi and collaboration with the FC and other environmental state institutions such as ZimParks and the Environmental Management Agency (EMA) enabled the WWF-Zimbabwe to obtain US$ 6.4 from the World Bank's Global Environment Fund (WWF 2014). These resources are not only critical for conservation in northwest Zimbabwe, but also for WWF-Zimbabwe's existence and expediency. NGOs often operate as non-profit organisations. Thus, funds obtained for conservation will also provide for the remuneration of staff, purchase of equipment, and infrastructure, hence supporting the existence of these organisations. WWF-Zimbabwe's interest in Sikumi Forest is, therefore, not just conservation but also the use of such frontiers as spatial spaces for fundraising.

Lastly, Sikumi Forest has additionally attracted private voluntary conservation organisations. An example is the Painted Dog Conservation (PDC), whose interest is in painted dog conservation (the dogs). The PDC owns land outside Sikumi Forest, but still relies on the reserve, first for painted dog conservation and second, for revenue generation. For conservation, Sikumi Forest provides an extended home range for dogs and their prey. The PDC fundraises in many forms, but one method that ties it to Sikumi Forest is fundraising through anti-poaching. The organisation runs an anti-poaching unit comprising civilian scouts. These scouts support the FC's anti-poaching efforts as part of PDC's broader wildlife conservation initiatives.[5] The data collected during anti-poaching activities is used to communicate success stories, financial needs, and challenges. The PDC does so through its tourist information centre situated along the road to Hwange national park, and through web 2.0 platforms such as Facebook. On the Internet, the PDC occasionally flashes "Did you know" facts on its website such as, "Did you know? Since 2001 we collected over 30,000 snares, enough to kill roughly 3000 animals."[6] Such information is advertised with the intention to market conservation activities in ways that advertise success, at the same time soliciting for funding. Together, this evidence demonstrates that NGOs are not only interested in frontiers like Sikumi Forests for conservation but also for economic gains.

Commercial Poachers

Commercial poachers are those illegally accessing resources from protected areas for large-scale commercial gains (Duffy and St John 2013). In Sikumi Forest, commercial poachers are interested in ivory poaching. They find Sikumi Forest an easy space to operate because while militarised guards securitise the forest area, they are not as sophisticated as park rangers from neighbouring Hwange national park. Sikumi Forest presents poachers with low risk and high returns. Another advantage for commercial poachers is that the forest is contiguous to local communities on its northeast borderline. Such proximity means that commercial poachers can be housed by local people and receive information regarding the movement of anti-poaching units and elephants. The reserve, therefore, presents commercial poachers with opportunities for successful ivory poaching ventures.

Local People

There are six communities surrounding Sikumi Forest. After being forcefully removed from the forest to its peripheries during the reservation era between 1930 and 1960, Sikumi Forest remains the only available source for supplementary livelihood resources for these communities. Population growth and the subsequent competition for resources have dwindled resources outside the forest. People either directly use these resources or sell them to supplement household income. What makes Sikumi Forest even more important for local communities is that the climate and soil in the region are not favourable for crop production. So, while agriculture is the primary livelihood source in most rural Zimbabwe, crop production for local people living adjacent to Sikumi Forest is on a small-scale basis and limited to drought-resistant crops such as sorghum and millet, which do not offer any diversity to the food basket. Livestock production is an alternative agriculture option but it is threatened by wildlife such as lions. Thus, for local people, Sikumi Forest is a safety net.

5. Divergent Values, Politics of Exclusion, and the Emergence of Violence

While Sikumi Forest converges a number of actors and interests, these interests are not in harmony. The state, NGOs, and safari operators have mutual interests in conservation in order to profit from it. This is unlike commercial poachers and local people who appear more interested in resource utilisation. For commercial poachers, illegally harvesting ivory is far more lucrative than worrying about conservation. It has been documented that there is greater incentive in taking the risk in armed poaching than in conserving elephants (Duffy and St John 2013; Moyle 2014; Krishnasamy et al. 2016; Spillane 2018). Commercial poachers are always ahead of state enforcement in their tact. In addition to using advanced automatic firearms, they also use lethal chemicals such as cyanide and operate in highly organised criminal syndicates.[7] State authorities and their conservation allies hence perceive commercial poachers' mode of resource utilisation and wealth accumulation as destructive and inconsistent with ideal conservation practice.

Local people's lack of concern for conservation is largely due to historical violent exclusionary practices and criminalisation of resource access. Communities living around Sikumi state that they do not benefit from forest conservation and ecotourism activities and are limited to less valuable resources such as firewood and thatch grass, which they must access through a restrictive and structurally violent permit system. Because of these restrictions they often resort to pilfering, setting fires in the process.[8] A concerned community member explained that it was not in the interest of community members to break the law, but restrictive and punitive conservation practices often led to non-compliance.[9] Other community members have gone as far as declaring a violence response to the FC's violent tactics.[10] Actions and threats of resistance are viewed by the FC as acts of sabotage to conservation programmes. The FC particularly perceives these people as a bunch of troublesome illegal encroachers good at pilfering, destroying infrastructure,

The Politics of Exclusion and Violence in Protected Areas 65

and causing forest fires (Forestry Commission 2015). It identifies people living adjacent to the forest as a category that does not value conservation.[11] Resource extraction activities by local people are, in similar ways to those of commercial poachers, seen as antagonistic to conservation efforts. There is, therefore, concerted determination to prevent illegal resource extraction for commercial or subsistence purposes. Efforts to halt illegal resource use are often confrontational and characterised by violence.

The state has always taken coercive and violent measures against what it perceives as illegal resource extraction. This violence is inherited and historical. It is embedded in a long history of territorialisation and fortress conservation practices that view people, particularly those living adjacent to parks, as threat (Peluso and Watts 2001; Brockington et al. 2008). The recent turn to militarisation in response to the poaching crisis has, in particular, witnessed increased intensity of violence as much as it has displayed the state's persistent violent protectionist and exclusionary character. The FC formally joined the collective national response to armed poaching in 1991 when it transformed its forest guards into a paramilitary guard force called the Forest Protection Unit (FPU).[12] Today, the FPU receives military training, uses military technology, and is characterised by excessive use of force. Force used by forest guards in Sikumi Forest ranges from shoot-to-kill tactics resulting in the death of suspects to physical, verbal, and psychological harassment and torture. Trespassers are assaulted, shoved around, and subjected to gruesome punishments. In some instances, trespassers may be killed. Forest guards also use derogatory and obscene language to interrogate and intimidate suspects during the arrest process.[13] Forest guards and FC officials insist that these enforcement strategies are necessary to maintain order in the forest and to save the remaining biodiversity.[14]

NGOs and private safari operators side with state-led conservation violence. Together, they maintain structural and physical barriers against users they perceive as a threat to their collective interest (Figure 4.2). The PPF, through the KAZA TFCA project, contributes to these barriers by funding and convening a law enforcement-working group, which includes the military and police from member states.[15] The inclusion of national security personnel, already known for their violent culture (Skjelsbaek 1979; Cock and Nathan 1989; Tendi 2013), increases the intensity of violence against resource users perceived as a threat, in symbolic and direct ways. But for the KAZA TFCA project, turning to militarised security personnel is, perhaps, a conservation strategy consistent with Article 5.1e of the KAZA treaty, which requires member states to prevent excessive utilisation of resources and to rehabilitate species faced with extinction (KAZA 2011). It could also be simply part of the broader PPF's ambitious vision of creating an untainted grand tourist-based economic zone, which in the process would also, however, contradict the very idea of peace and inclusivity enshrined in the Peace Parks concept (Büscher and Ramutsindela 2016).

66 Tafadzwa Mushonga

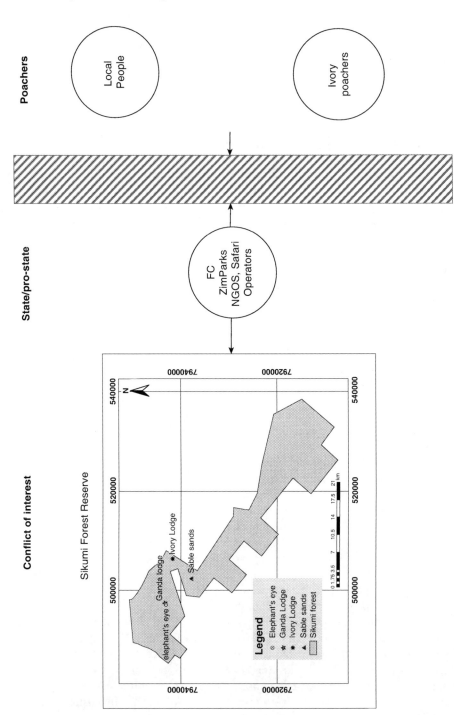

Figure 4.2 Structural and physical barriers in resource extraction.

WWF-Zimbabwe has strengthened state capacity by supporting militarised anti-poaching operations. Using funding obtained from the World Bank, the organisation officially handed over military equipment to its state partners. The FC received two 4-wheel drive vehicles, webbing jackets, water bottles, binoculars, and installation of Spatial Monitoring and Reporting Tool (SMART) technology and a radio communication system.[16] In 2017, WWF funded advanced military training for forest guards in Sikumi. The irony of WWF-Zimbabwe's participation in these violent exclusionary efforts is that the HSBCP is marketed as a community-participation-centred project. Yet, the WWF is funding the violent placing of the very same people at the peripheries of conservation and resource access. The role played by the WWF in Sikumi Forest confirms the argument that people are often used for the purposes of generating funding and neglected in favour of state goals (Brockington et al. 2008; Benjaminsen and Bryceson 2012). It also displays the nature of politics surrounding exclusion in resource extraction.

The Painted Dog Conservation, on the other hand, assists the FC with everyday policing operations. In addition to increasing boots on the ground, the organisation perpetuates violent exclusionary practices by incentivising anti-poaching activities. It pays a poacher's bonus for every successful arrest process.[17] Private safari operators have supported these anti-poaching incentives by pledging to pay amounts as high as US$ 500 per team for every successful anti-poaching operation.[18] As highlighted earlier, these efforts are undertaken with specific interests inclined towards monetary gains from conservation. However, they translate to increased violence. A poacher's bonus – money paid to the anti-poaching team for successfully apprehending an illegal access activity in the forest, is only paid when there is sufficient evidence that the apprehended suspect was indeed involved in illegal activities. Such terms and conditions often mean that forest guards use violence to obtain such evidence. A common method in Sikumi Forest involves subjecting suspects to various forms of physical and verbal harassment, coercing them to submit exhibits, and plead guilty to allegations.

Overall, collective efforts of the state, and its conservation allies are all set to create a "securitized green spatial fix for the overaccumulation of capital" (Massé and Lunstrum 2016: 228) through which only this category of users can benefit by violently excluding others in structural and direct ways. There is copious literature showing that overaccumulation of capital and violent practices are disguised in justifications for conservation and "win–win" gains. I do not dispute the importance of conservation, even the possibility of win–win benefits. However, like others who criticise the militarisation of conservation, I am concerned that the increasing reliance on extreme forms of violence to territorialise resource-rich frontiers and to exclude already disenfranchised people regress the gains and expected future outcomes of conservation. In Sikumi Forest, it is local people who largely experience state-led violent exclusionary mechanisms. Unlike commercial poachers who, in addition to being highly organised and sophisticated in their tact use sophisticated automatic weaponry, the *modus operandi* of most local people is less complex, for subsistence and often involving the use of livelihood and safety tools such as catapults, axes, sometimes accompanied by dogs. This makes local people easy to police and more susceptible to state violent tactics. Thus, the following discussion on the implications of violent exclusionary conservation tactics is more inclined towards experiences of local communities.

6. Violent Exclusion and Implications for Conservation

Violent exclusion, as it unfolds in frontiers like Sikumi Forest, is often justified as part of broader ideal practices to save the remaining biodiversity from extinction (Terborgh 2004; Brockington et al. 2008). It has, however, been argued that violent conservation practices such as militarised anti-poaching activities witnessed in Sikumi Forest could, in fact, undermine long-term conservation goals (Duffy et al. 2015). I locate my discussion within this perspective but touch on implications for conservation from a human rights and environmental justice lens. In doing so, I do not claim to be a voice for local people or be an advocate of human rights. I rather wish to draw attention to the corollary effects of violent exclusionary resource extraction in ways that enable us to engage with the future of conservation in violent environments.

Thus far, I have shown how resource extraction practices used in resource-rich protected areas privilege a few elites, how they are founded on logics of violent exclusion, and how such violence unravels. Direct violence such as that perpetrated by paramilitary forest guards, supported by the state and its conservation allies, has immediate effects on physical, verbal, and psychological abuse of local people. However, the structural effects of exclusion have far-reaching consequences on distributional, and procedural justice, and overall people's rights. For example, violent exclusion from resource-based entrepreneurships means that ordinary people are unable to partake in resource-based economic development. Other spheres of people's development reliant on economic stability, such as health, shelter, and education, which according to international and national law are basic human rights,[19] also become compromised in slow and corrosive ways (Nixon 2011; Dressler and Guieb III 2015).

Slow violence and the corrosive nature of exclusionary resource extraction also often have a bearing on people's traditions and culture (Survival International 2014; Thondhlana and Shackleton 2015). For example, forest gathering is a long-existing cultural practice from which indigenous knowledge has been accumulated, yet, it is upon the basis of such a practice that protected areas have become territories of coercive and exclusionary management (Steinhart 1989; Kwashirai 2009). Galtung (1990) has argued that when the state uses cultural practices to rationalise violent exclusion from resources, they subject local people to cultural violence. Within the context of cultural violence, using conservation to justify violent exclusionary approaches deprives resource-dependent people from exercising and maintaining ways of life linked to resource access, and their right to preserve culture and tradition defining their very being. These examples are by no means exhaustive, but they serve to demonstrate the linkages between exclusionary violent resource extraction, and environmental injustices in and around protected areas.

There are also implications for conservation. Duffy (2014) argues that the adoption of violent conservation practices alienates local communities who could potentially help with successful conservation outcomes. Survival International, a global movement for tribal people's rights since 1969, has also argued that local people are the "eyes and ears" of the forest, explaining that "those who rely on their land to survive are more motivated to protect their environment than poorly paid forest guards" (Survival International 2014: 17). Using case examples from central, east, and southern Africa and parts of Asia, Survival International has further demonstrated how violent evictions and policing practices in protected areas produce disgruntled people who then go out to undermine conservation initiatives. This has also been observed in Sikumi Forest. The violent exclusion of local people from resource access and consequent violation of fundamental human rights has produced a characteristically resistant and violent category whose activities have become a deliberate source of environmental degradation within the reserve, in defiance of state-led violent practices. Signs of deliberate sabotage of conservation efforts are already manifesting in Sikumi Forest in covert and overt ways. Like in other forest reserves in Zimbabwe, arson, pilfering, and confrontations with forest officers and guards are not uncommon (Mapedza and Mandondo 2002; Matose 2002; Mushonga 2018).

Thus, a significant implication of human rights violations and environmental injustice is that disgruntled people will less likely support conservation initiatives (also see, Duffy and St John 2013; Annecke and Masubelele 2016; Hubschle 2017). In one of the villages adjacent to Sikumi Forest, an anonymous man, whose views were echoed by several community members, explained how they often see and leave the forest burning. He explained,

> We know that the forest guards are few. They cannot not look after the whole forest or control fire by themselves. They would need us in our numbers. But with the way they treat us it is better to watch the forest burning. It is possible that if you try to assist forest guards, they can turn around to accuse you of stealing resources from the forest and abuse you.[20]

A village head from one of the adjacent villages explained how community leaders are willing to help with curbing ivory poaching. They have undertaken a vetting exercise on all new people in their villages in a bid to show their contribution towards halting ivory and timber poaching and to perhaps change the perception of the FC about local people. However, the community's desire to assist with conservation has not stopped the violent treatment and othering of local people, thereby frustrating their efforts.[21] So, while exclusionary approaches are defended as one of the only realistic ways to protect biodiversity (see Hutton et al. 2005; Büscher and Fletcher 2018), such practices cannot be discussed in the context of conservation alone. Corollary effects, particularly on resource-dependent people, demand that such practices be discussed within the wider context of human rights and the effect of violating such rights on conservation outcomes.

7. Concluding Remarks

The politics of exclusion and violence in nature reserves occurs against the backdrop of territorialisation, securitisation, and militarisation of conservation, and is perpetuated by converging but conflicting resource extraction interests. I have presented how the politics of resource use and exclusion unfolds. I have also shown how conservation is used to justify selfish neoliberal ends, and by violent means. The intention of this chapter is not to dispute the value of conservation, but the motives of such a cause, as well as the approaches for achieving such motives. Clearly, drivers for conservation are, today, increasingly financially motivated than they are driven by the desire to protect biodiversity, the outcome of which sustains violent exclusion and othering in resource-rich frontiers. The consequences of environmental injustice as a result of violent exclusionary conservation practices have counterproductive costs on intended biodiversity protection. To reverse such costs, I sugges that violent exclusionary resource extraction practices be engaged beyond the rubric of conservation science to a broader humanities context on human rights. Given the wide-ranging connection between human rights, resource extraction, and overall environmental health, such an approach could offer wider engagement with the politics of exclusion and violence in resource-rich geographies.

Notes

1 Interview, ZimParks official 05/08/2016.
2 www.forestry.co.zw.
3 Zimparks.org.
4 Forestry Commission internal memorandum dated 02/2016.
5 PDC anti-poaching official 05/2016.
6 www.painteddog.org/education-and-outreach-programs/.
7 Group discussion with forest guards 05/2016; ZimParks official 08/2016.
8 Community meeting at Mabale Village 06/2016.
9 Anonymous community member from Dopota village 06/2016.
10 Anonymous community members from Jwapi Village 07/2016.
11 Personal working experiences with the FC since 2007.
12 FC's Chief security officer 04/2016.

13 Interviews with local people 07–11/2016; field observations.
14 Forest guards, FC officials and local communities, as well participant observation of anti-poaching activities.
15 Interview, ZimParks official 08/2016.
16 Handover over ceremony of HSBCP equipment to project partners. The ceremony was held at the WWF-Zimbabwe Harare offices on 17 December 2015.
17 Interview, PDC anti-poaching scout 04/2016.
18 Forestry Commission internal memorandum, 02/2016.
19 For example, the United Nations Declaration on the Rights of Indigenous Peoples and national constitutions.
20 Interview, anonymous man.
21 Interview, anonymous village head.

References

Andersson, J. A. et al. 2013. TFCAs and the invisible peoples. In *Transfrontier Conservation Area: People Living on the Edge*, ed. J. A. Anderson, M. D. Garine-Wichatitsky, D. H. M. Cumming, V. Dzingirai and K. Giller. London and New York: Routledge.

Annecke, W. and M. Masubelele. 2016. A review of the impact of militarisation: The case of rhino poaching in Kruger National Park, South Africa. *Conservation and Society* 14:195–204.

Benjaminsen, T. A. and I. Bryceson. 2012. Conservation, green/blue grabbing and accumulation by dispossession in Tanzania. *Journal of Peasant Studies* 39:335–355.

Bernazzoli, R. M. and C. Flint. 2010. Embodying the garrison state? Everyday geographies of militarization in American society. *Political Geography* 29:157–166.

Bocarejo, D. and D. Ojeda. 2016. Violence and conservation: Beyond unintended consequences and unfortunate coincidences. *Geoforum* 69:176–183.

Bradley, E. and K. McNamara. 1993. *Living With Trees: Policies for Forestry Management in Zimbabwe*. Washington, DC: World Bank Group.

Brockington, D. and J. Igoe. 2006. Eviction for conservation: A global overview. *Conservation and Society* 4:424–470.

Brockington, D. et al. 2006. Conservation, human rights, and poverty reduction. *Conservation Biology* 20:250–252.

Brockington, D. et al. 2008. *Nature Unbound: Conservation, Capitalism and the Future of Protected Areas*. London: Earthscan.

Büscher, B. and R. Fletcher. 2018. Under pressure: Conceptualising political ecologies of green wars. *Conservation and Society* 16:105–113.

Büscher, B. and M. Ramutsindela. 2016. Green violence: Rhino poaching and the war to save southern Africa's peace parks. *African Affair* 115:1–22.

Büscher, B. et al. 2014. *Nature Inc.: Environmental Conservation in the Neoliberal Age*. Tucson, AZ: University of Arizona Press.

Cock, J. and L. Nathan. 1989. *War and Society: The Militarisation of South Africa*. Cape Town, South Africa: New Africa Books.

Dowie, M. 2005. Conservation refugees. *Orion* 24:16–27.

Dressler, W. H. and E. R. Guieb III. 2015. Violent enclosures, violated livelihoods: Environmental and military territoriality in a Philippine frontier. *Journal of Peasant Studies* 42:323–345.

Dudley, N. and S. Stolton. 2010. *Arguments for Protected Areas: Multiple Benefits for Conservation and Use*. New York: Routledge.

Duffy, R. 2010. *Nature Crime: How We're Getting Conservation Wrong*. New Haven: Yale University Press.
Duffy, R. 2014. Are we hearing a 'call to arms' from wildlife conservationists? *https:// justconservation.org/are-we-hearing-a-call-to-arms* (accessed July 20, 2017).
Duffy, R. 2016. War, by conservation. *Geoforum* 69:238–248.
Duffy, R. 2017. We need to talk about militarisation of conservation. *www.greeneuropeanjournal.eu/we-need-to-talk-about-militarisation-of-conservation/* (accessed August 25, 2017).
Duffy, R. and F. St John. 2013. *Poverty, Poaching and Trafficking: What Are the Links?* DOI: http://dx.doi.org/10.12774/eod_hd059.jun2013.duffy
Duffy, R. et al. 2015. The militarization of anti-poaching: Undermining long term goals? *Environmental Conservation* 42:345–348.
Duffy, R. et al. 2019. Why we must question the militarisation of conservation. *Biological Conservation* 232:66–73.
Fairhead, J. et al. 2012. Green grabbing: A new appropriation of nature? *Journal of Peasant Studies* 39:237–261.
Feld, M. D. 1977. *The Structure of Violence: Armed Forces as Social Systems*. Beverly Hills, CA: Sage Publications, Inc.
Fisher, W. F. 1997. Doing good? The politics and antipolitics of NGO practices. *Annual review of Anthropology* 26:439–464.
Forestry Commission. 2013. Position paper on illegal settlement in all the gazetted forests of Zimbabwe. Unpublished. Forestry Commission.
Forestry Commission. 2015. *Tracking Tool for Biodiversity Projects in GEF-3, GEF-4, and GEF-5: Management Effectiveness Tracking Tool (METT): Sikumi and Ngamo Forests*. Harare: Forestry Commission.
Galtung, J. 1990. Cultural violence. *Journal of peace research* 27:291–305.
Given, M. 2002. Maps, fields, and boundary cairns: Demarcation and resistance in colonial Cyprus. *International Journal of Historical Archaeology* 6:1–22.
Guha, R. 2000. *The Unquiet Woods: Ecological Change and Peasant Resistance in the Himalaya*. Berkeley, CA: University of California Press.
Holmes, G. 2014. Defining the forest, defending the forest: Political ecology, territoriality, and resistance to a protected area in the Dominican Republic. *Geoforum* 53:1–10.
Hubschle, A. M. 2017. The social economy of rhino poaching: Of economic freedom fighters, professional hunters and marginalized local people. *Current Sociology* 65:427–447.
Hutton, J., W. M. Adams and J. C. Murombedzi. 2005. Back to the barriers? Changing narratives in biodiversity conservation. *Forum for Development Studies* 32:341–370.
Kavango-Zambezi Organisation. 2014. KAZA TFCA to launch a KAZA VISA Pilot project between the Republic of Zambia and Zimbabwe on 28 November 2014. *www.kavangozambezi.org/kaza-tfca-launch-kaza-visa-pilot-project-between-republic-zambia-and-zimbabwe-28th-november-2014* (accessed March 31, 2017).
KAZA. 2011. Kavango Zambezi trans frontier conservation area treaty. *https://tfca-portal.org/kaza-tfca-treaty* (accessed March 31, 2017).
Kelly, A. B. and M. Ybarra. 2016. Introduction to themed issue: "Green security in protected areas". *Geoforum* 69:171–175.
Krishnasamy, K. et al. 2016. In transition: Bangkok's ivory market. In *An 18-month Survey of Bangkok's Ivory Market*. Petaling Jaya: TRAFFIC Southeast Asia.

Kwashirai, V. 2009. *Green Colonialism in Zimbabwe, 1890–1980.* New York: Cambria Press.

Laterza, V. and J. Sharp. 2017. Extraction and beyond: People's economic responses to restructuring in southern and central Africa. *Review of African Political Economy* 44:173–188.

Lewis, M. L. 2010. Conservation refugees: The hundred-year conflict between global conservation and native peoples (review). *Global Environmental Politics* 10:122–124.

Lunstrum, E. 2014. Green militarization: Anti-poaching efforts and the spatial contours of Kruger National Park. *Annals of the Association of American Geographers* 104:816–832.

Lunstrum, E. and M. Ybarra. 2018. Deploying difference: Security threat narratives and state displacement from protected areas. *Conservation and Society* 16:114–124.

Mapedza, E. and A. Mandondo. 2002. *Co-management in the Mafungautsi State Forest Area of Zimbabwe: What Stake for Local Communities?* Washington, DC: World Resources Institute.

Massé, F. and E. Lunstrum. 2016. Accumulation by securitization: Commercial poaching, neoliberal conservation, and the creation of new wildlife frontiers. *Geoforum* 69:227–237.

Matose, F. 1997. Conflicts around forest reserves in Zimbabwe: What prospects for community management? *IDS Bulletin* 28:69–78.

Matose, F. 2002. Local people and reserved forests in Zimbabwe: What prospects for co-management? PhD diss., University of Sussex.

Matose, F. 2014. Nature, villagers, and the state: Resistance politics from protected areas in Zimbabwe. In *Nature Inc: Environmental Conservation in the Neoliberal Age*, ed. B. Buscher, W. Dressler, and R. Fletcher. Tucson, AZ: University of Arizona Press.

Moyle, B. 2014. The raw and the carved: Shipping costs and ivory smuggling. *Ecological Economics* 107:259–265.

Mudekwe, J. 2007. The impact of subsistence use of forest products and the dynamics of harvested woody species populations in a protected forest reserve in western Zimbabwe. PhD diss., Stellenbosch University.

Mushonga, T. 2018. The militarisation of conservation, violence and local people: The case of Sikumi forest reserve in Zimbabwe. PhD diss., University of Cape Town.

Neumann, R. P. 2004. Moral and discursive geographies in the war for biodiversity in Africa. *Political Geography* 23:813–837.

Nixon, R. 2011. *Slow Violence and the Environmentalism of the Poor.* Cambridge, MA: Harvard University Press.

Ojeda, D. 2012. Green pretexts: Ecotourism, neoliberal conservation and land grabbing in Tayrona National Natural Park, Colombia. *Journal of Peasant Studies* 39:357–375.

Peluso, N. L. and P. Vandergeest. 2001. Genealogies of the political forest and customary rights in Indonesia, Malaysia, and Thailand. *The Journal of Asian Studies* 60:761–812.

Peluso, N. L. and M. Watts. 2001. *Violent Environments.* Ithaca, NY: Cornell University Press.

Skjelsbaek, K. 1979. Militarism, its dimensions and corollaries: An attempt at conceptual clarification. *Journal of Peace Research* 16:213–229.

Spierenburg, M. et al. 2006. Resistance of local communities against marginalization in the great Limpopo transfrontier conservation area. *Focaal* 18–31.

Spillane, J. J. 2018. Poaching: A moral issue and a failure of the market. *La Civiltà Cattolica, English Edition* 2:21–30.

Steinhart, E. I. 1989. Hunters, poachers and gamekeepers: Towards a social history of hunting in colonial Kenya. *The Journal of African History* 30:47–264.

Survival International. 2014. Parks need people. *www.survivalinternational.org* (accessed March 31, 2017).

Tendi, B.-M. 2013. Ideology, civilian authority and the Zimbabwean military. *Journal of Southern African Studies* 39:829–843.

Terborgh, J. 2004. *Requiem for Nature*. Washington, DC: Island Press.

Thondhlana, G. and S. Shackleton. 2015. Cultural values of natural resources among the San people neighbouring Kgalagadi Transfrontier Park, South Africa. *Local Environment* 20:18–33.

Vandergeest, P. and N. L. Peluso. 1995. Territorialization and state power in Thailand. *Theory and Society* 24:385–426.

West, P. et al. 2006. Parks and peoples: The social impact of protected areas. *The Annual Review of Anthropology* 35:251–277.

Wilkie, D. S. et al. 2006. Parks and people: Assessing the human welfare effects of establishing protected areas for biodiversity conservation. *Conservation Biology* 20:247–249.

WWF. 2014. HSBC project. *https://wwf.panda.org/wwf_offices/zimbabwe/our_work/hsbc_project/* (accessed February 2, 2015).

5 The Politics of Mining Pollution in Zambia

Investigating 100 Years of Environmental Management on the Copperbelt

Chibamba Jennifer Chansa

Introduction

Large-scale copper mining activities in Zambia are currently concentrated in the Copperbelt region.[1] Despite its long existence, early accounts on the region did not address environmental issues. The closest that early accounts on the region came to environmental studies were through geological studies such as those by Jackson (1933) as well as Gray and Parker (1929), which provided the basis for later environmental histories of the region. More recent studies that address environmental transitions and practices include those by Schumaker (2008) and Ross (2017), which highlight the impact of disease control during colonial rule on the environment and provide insight into early environmental management strategies.

The long existence of commercial mining on the Zambian Copperbelt and its significant impact on the environment warrants an extensive study of the environmental history of the region, as mining inevitably contributes to environmental degradation by altering the earth's surface and surrounding environment (Blaikie and Brookfield 2015: 1). In addition to environmental contamination, mining produces excessive tailings, harmful gaseous emissions, and solid waste that are difficult to dispose of. Its establishment, therefore, creates conditions for environmental pollution (Styve 2013: 18).

Although a few studies on the Zambian Copperbelt have addressed environmental issues, none has provided a detailed trajectory of developments in environmental management and regulation since colonial rule. This chapter highlights the complexity in state-mining company interactions that have a significant impact on mining investment, as well as mineral and environmental governance. An examination of the industry over what has been almost one decade of mining exploitation is significant, given various transitions in mine ownership, governance, and regulation since the 1920s.

Drawing heavily from research conducted on the Zambian Copperbelt between 2017 and 2018, this chapter broadly examines the evolution of environmental concerns in the region since the 1920s, in line with local developments. It specifically highlights the impact of independence, nationalisation, privatisation, and other political transitions on the regulation of the mining industry, and the consequences on environmental management. By so doing, it contributes to scholarship on the link between local economic and political developments in Africa, and their impact on natural resource and environmental governance.

Mining, Health, and Safety on the Zambian Copperbelt: The First 40 Years

Although commercial mining in Zambia was first commissioned at Luanshya in 1928, privately managed mining activities began long before that time (Lungu 2008a: 544, 2008b: 404). When the British Colonial Office took over the governance of Northern Rhodesia in 1924, the mineral rights previously owned by the British South Africa Company (BSAC), which ruled the territory between 1890 and 1924, were not transferred to the Crown but remained under the Company (Adam and Simpasa 2011: 308; Hansungule et al. 1998: 17). The South African mining firm Anglo American Corporation (AAC) and Alfred Chester Beatty's Rhodesian Selection Trust (RST) operated commercial mining operations on the Copperbelt from the late 1920s until independence (Muchimba 2010: 5). RST eventually owned five mines on the Copperbelt, which included Mufulira, Luanshya, Chibuluma, Chambishi, and Kalengwa. Of these, four were underground mines. Facilities at some of the RST mines included a concentrator, smelter, and refinery. AAC owned the Nkana (then Rhokana) and Nchanga mines. Facilities at the mines also included a concentrator, smelter, and refining plant, as well as roaster leach facilities that enabled the recycling of copper anode slimes (Sikamo et al. 2016: 492). Already, the presence of these facilities created concerns for various forms of pollution and environmental contamination.

The outbreak of the Second World War in 1939 resulted in increased mineral production in mineral-rich colonies in Africa but had negative consequences on the health and safety of mineworkers, as well as on environmental safety. Copper demand increased during the wartime years due to its usage in the production of ammunition and aircrafts (Dumett 1985: 393). Increased copper production in Northern Rhodesia led to several developments including extraction of ores at greater depths, as well as expansion of mining facilities at various mines (National Archives of Zambia ML8-11-99 1955).

Mining expansion presented three major disadvantages. Firstly, mining at deeper levels increased the potential for mineworkers to contract dust-related diseases such as silicosis (NAZ ML8-11-99 1955). Secondly, some new mining techniques increased the health and safety risks of mineworkers by exposing them to more hazardous conditions than before. Among these techniques was open-pit mining, introduced during the 1950s, which enabled the extraction of orebodies that were located close to the surface. Although the technique was generally considered less hazardous than underground mining, it contributed to land degradation, dust, noise, and possible altering of groundwater flow. The contaminated water was unsuitable for human and animal consumption and threatened the survival of living organisms in and around the affected areas (Monjezi et al. 2009: 206). The impact of open-pit mining on human life, agriculture, and ecosystems meant that it was just as unsafe as underground mining. Thirdly, mining expansion created challenges in waste management. By the late 1950s, the Copperbelt mines jointly produced an estimated 1,750,000 tonnes of tailings monthly.[2] Mining companies, therefore, turned to cheaper waste disposal methods such as the use of naturally existing valleys as disposal sites, whose establishment and maintenance cost less than environmentally sound disposal methods. Furthermore, the establishment of natural tailings deposits presented opportunities for recreational activities such as boating. For this reason, little regard was given to the potentially hazardous consequences of using natural valleys for tailings disposal.

The waste disposal method remained popular on the Copperbelt well into the post-independence period. Its popularity does not only demonstrate the preference of mining companies for cost-effectiveness at the expense of environmental safety, but also the inability of the colonial government to regulate mining activities effectively. This was demonstrated by the colonial government's inability to outlaw the practice despite the occurrence of accidents that involved the collapse of tailings-filled natural valleys as early as 1952.[3]

The popularity of this waste disposal method on the Copperbelt further demonstrates negligible waste management on the Copperbelt. In fact, waste contributed even further to the environmental decline of the Copperbelt when it began to be used to dry out mosquito-infested swamps as a means of malaria control from the late 1920s (Ross 2017: 179; Schumaker 2008: 826–828). While the practice was successful in reducing malaria prevalence and infection rates among mineworkers, it left many swamps on the Copperbelt void of flora and fauna. This was because of the toxicity of the tailings. Therefore, waste management strategies introduced during the colonial period resolved immediate disposal and health concerns but contributed to the gradual degradation of the environment.

Despite the limited attention to the environmental consequences of mining, colonial government, and mining officials were in fact aware of some of the impacts of mining. Even then, environmental management strategies were either limited to the production of specific products or addressed environmental issues ineffectively. For example, demands were made by colonial authorities for relevant legislation to address the safety of workers in the uranium industry and those working in open-pit mines in Northern Rhodesia in the mid-1950s.[4] These concerns, however, did not apply to workers engaged in sulphur and sulphuric acid production on several Copperbelt mines during that same period.[5] Concerns for the uranium industry demonstrate that mining and government officials were not entirely oblivious to the effects of mining on the environment, as well as the health and safety of employees. On the other hand, the inability to extend similar provisions to the sulphur industry validates the fact that some aspects of mining were simply neglected.

The enactment of mining laws in Northern Rhodesia was generally problematic. This was due to the 'protectorate' status of the country at that time, which required effective collaboration between the British Colonial Office and the Northern Rhodesia government before laws could be enacted. Furthermore, the influence of the BSAC and mining companies in law-making had a negative impact on the enactment of certain proposed laws. Given the colonial politics of the time, mining laws often safeguarded the interests of European employees at the expense of African workers. This limited the impact of enacted laws, especially given that African mineworkers were greater in number than European workers were. For example, European mineworkers received higher compensation for industrial diseases than Africans did, despite working in the same industry with exposure to similar hazards.[6]

Another disadvantage of mining law in Northern Rhodesia was that it generally developed at a slow pace in comparison to other mining countries. For example, in the United States of America, legislation addressing air pollution was first enacted in 1881. At that time, the major success of the legislation was that it classified smoke as a public nuisance, thereby introducing liabilities for offenders. However, by 1890, the law advanced further by prohibiting the emission of substances beyond specified limits without prior notice. The legislation then evolved to focus on the prevention of hazardous air pollutants. In 1955, the first comprehensive federal air pollution law was enacted (Stern 1982: 44). Although the enactment of effective legislation in the USA took several decades, concerns for air pollution gradually transitioned into more prevention-specific laws targeting hazardous emissions. This was unlike the case in Northern Rhodesia, where insufficient laws that facilitated pollution by mining companies lasted many decades without transitioning into effective laws that prevented pollution.

The legislation addressing mining-related pollution that was eventually enacted in Northern Rhodesia provided indemnity for mining companies from prosecution for environmental crimes. This was particularly the case for legislation addressing air pollution. For example, the Smoke Damage (Prohibition) Act of 1935 protected declared mining areas on the Copperbelt by categorising them as "smoke areas." Therefore, mining companies could not be prosecuted for pollution resulting from smelter emissions including sulphur dioxide (Mwaanga et al. 2019: 3). This facilitated increased pollution, as there were no consequences for emitting hazardous substances. In many ways, colonial legislation laid the foundation for future practices that would continue to favour mining companies at the expense of environmental safety.

By the 1950s, an increased understanding of industrial diseases contributed to the enactment of various health and safety laws, and revisions to existing ones, that helped address the challenges previously faced in managing industrial diseases and safety (Paul 1961: 96–109). However, the focus on the health and safety of mineworkers meant that very limited attention was given to the impact of mining on the actual environment, as the two aspects were treated almost separately. By limiting attention and legislation to industrial diseases, first aid, and safety while within the mine; the effects of mining pollution that were in fact the cause of some mining-related health conditions remained poorly addressed.

Mining Regulation During UNIP Rule

It was against this background that the United National Independence Party (UNIPs) took over the running of the industry in 1964. The UNIP Government formulated the ideology of Humanism that, according to President Kaunda, placed man at the centre of God's creation and in command of social, economic, and political factors. According to the UNIP government, the philosophy was created to enable Zambians to exercise control over the means of production in order to facilitate wealth distribution. However, the philosophy ultimately justified the government's efforts to enhance its authority and control (Kanu 2014: 376–377; Shaw 1976a: 79).

Given the existing ownership structure of the mining industry at the time of independence, the Government's primary focus in terms of legislation was to secure mineral ownership (Saasa 1985: 359). Soon thereafter, the focus turned to the nationalisation of the industry. The 1969 Mines and Minerals Act facilitated the process (Ndulo 1986: 11). Among the reasons for the nationalisation of the industry was the alleged lack of reinvestment into the industry by mining companies. Mining officials denied this, claiming that their failure to reinvest through the payment of taxes was due to their high expenditure on environmentally sound mining practices (Saasa 1985: 361). Contrary to their claims, mining companies were in fact contributing to environmental degradation through cost-effective but environmentally hazardous mining techniques and disposal methods. For example, the use of naturally existing valleys for tailings disposal, introduced during the 1930s, was still in use at several mines by the time of nationalisation. This contradicts mining company claims about the use of costly environmentally sound mining practices in the industry and demonstrates the fact that they occasionally downplayed mining pollution to defend their position with the state. Furthermore, the presence of hazardous mining practices by the 1960s highlights the fact that the UNIP government inherited a mining industry that was already faced with significant environmental challenges.

Through nationalisation, the Government obtained 51% ownership of AAC and RST's operations, as well as in any of their new projects. This meant that the Government held controlling rights over the industry, while the mining companies retained 49% of the shares (Bebbington et al. 2018: 121). By December 1969, Nchanga Consolidated Copper Mines (NCCM) and Roan Consolidated Mines (RCM) operated the Copperbelt mines (Saasa 1985: 364).

The ownership structure of the mining industry that emerged after nationalisation had an adverse impact on environmental management. This was because, despite the states' controlling rights over the industry, the Government was in fact not effectively involved in the actual operations of mining companies. In addition to this, the Articles of Association between the two parties granted the mining companies management and decision-making rights that left them in control of the industry (Saasa 1985: 364–365). The implication of this was that the running and practices on the mines remained the same despite nationalisation. Mining companies, therefore, continued to focus on the impact of mining on health and safety, while limited attention was paid to mining-related environmental concerns. Even then, efforts remained limited to safety "within" the mines rather than the wider environment and mineworkers once they left the mine premises. These were equally significant areas, given the extensive nature of some forms of pollution. Furthermore, pollution outside the mines was possible through incidents such as accidents during the transportation of hazardous substances to offsite locations and subsequent contamination of the areas surrounding these 'outside' locations. In addition to this, mine facilities were often spread out across a town, creating additional safety concerns. The control that mining companies retained over operations, therefore, limited any chances for improved environmental regulation.

The continued reliance on health, safety, and environmental measures introduced during the colonial period was detrimental for the industry and the economy. For example, the continued use of natural valleys as tailings deposits resulted in a tragic collapse of a tailings dam at Mufulira Mine in 1970 (Vutukuri and Singh 1995: 120). The accident, which was similar but more tragic than that in 1952, revealed additional weaknesses in reporting within the mines, limited mine inspections, and questioned the decision-making of the Mines Department that approved the establishment of mine structures (Government of the Republic of Zambia 1971: 18). The tragedy resulted in the destruction of mining infrastructure and equipment, leading to reduced average copper production at the mine from 617,000 tonnes per month to only 30,000 tonnes by the end of 1970. This was an enormous setback, given the projections for April 1971, which had predicted copper output at 324,000 tonnes. In addition to the financial costs of rehabilitating the mine, additional costs were accrued in the compensation of the families of deceased miners (Vutukuri and Singh 1995: 120; Tembo 2019: 126). The 1970 accident, therefore, highlighted the gravity of poor environmental management on the mines and poor regulation by government departments.

Despite the significance of the 1970 accident, limited changes occurred in the regulation of the industry. In terms of legislation, the only significant changes aimed at preventing incidents similar to that in 1970 were provisions in the 1971 and 1973 Mining Regulations that addressed surface protection within mining sites.[7] To some extent, these provisions strengthened mine safety. Furthermore, the Ministry of Mines began implementing the recommendations of the Commission of Inquiry into the 1970 accident soon after its occurrence, which included improvements in waste dumping regulations (Tembo 2019: 127–128). Still, the lack of significant reform in environmental legislation meant that the changes that occurred in mining regulation during the 1970s had minimal impact. This was worsened by economic challenges such as the oil and copper crises of the 1970s, which limited the state's ability to fund the industry, let alone focus on environmental safety (Sill 2007: 22; Burdette 1984: 208–209; Shaw 1976b: 14).

Locally, the state's ability to focus on environmental protection was also affected by internal politics within UNIP. In 1972, the party introduced one-party rule to secure its political position as the ruling party (Larmer 2016: 4–5). The following year, the state also terminated all existing mining contracts with the private shareholders in NCCM and RCM (Saasa 1985: 364–365). While these actions secured UNIP's position over the country and mining industry, they negatively affected the party's ability to introduce regulatory reform. Already, the financial challenges within the country meant that the industry was poorly financed. Within the party, the shift to one-party rule created internal conflict. This was because President Kaunda and his political advisors took on the responsibility of policymaking in the country, while government officials were excluded from the process. This initially worked in favour of the President when he had the full support of the party and parliament (Gibson 1994). This was not the case, however, when it came to the enactment of natural resource conservation laws. Kaunda lost support from party officials when they began to feel that proposed laws were restricting their personal access to natural resources (Demers 2015). Therefore, despite President Kaunda's passion for natural resource conservation, opposition to the policy-making process within UNIP limited the government's ability to protect the environment through legislative reform.

Unfortunately, the political opposition and financial constraints coincided with the emergence of international concerns for natural resource protection, which would have otherwise benefitted the country. Developments at that time included the United Nations Conference on the Human Environment in 1972, which discussed the need to manage the human environment effectively, and the World Meteorological Organisation technical conference in 1973 on the observation and measurement of atmospheric pollution. The technical conference addressed many significant aspects that were important for mining countries like Zambia. For example, the conference highlighted the requirements for the measurement of air pollution, air quality management, and the transportation of air pollutants (World Meteorological Organization 1973). The emphasis on air pollution was particularly important given the effects of mining on air pollution on the Copperbelt. However, the state did not urgently address mining-induced air pollution in line with the recommendations of the conference. By the 1980s, global environmental concerns were given even greater significance. The World

Commission on Environment and Development and the Brundtland Commission of 1984, as well as the Basel Convention of 1989, demonstrated this (United Nations 1987; Krueger 2002: 43–44; Martinez-Alier 2001: 18). Environmental services were initially extended to the African continent through the United Nations Environment Program's establishment of an office in Kenya in 1984.[8]

The developments of the 1970s and 1980s did not result in particularly significant changes to natural resources and environmental measures in Zambia, particularly in the mining industry. Although the late development of global natural resource concerns partly explains the slow development of local efforts, the technical conferences and events of that time marked an 'increase' rather than 'genesis' in natural resource protection efforts. Therefore, even in developing resource-rich countries like Zambia, the developments of the 1970s and 1980s simply emphasised concerns that were already known within the country and industry. To the detriment of third-world countries, however, the developments – particularly those of the 1980s – focused on 'sustainable development' at a time when newly independent poorer countries were only just beginning to define economic development on their own terms (Chambers and Gordon 1991). Although this inevitably meant that the developments were unsuitable for less developed countries like Zambia, the limited approach to environmental protection was also the result of the local negligence of growing environmental challenges on the Copperbelt. The economic and political factors experienced after independence similarly did not fully justify the state's inability to regulate the industry and environment effectively. This is justified by the state's failure to ensure environmental safety even during more profitable periods (Simpasa et al. 2013: 5–6). This suggests that although the slow development of global concerns and political and economic challenges contributed to limited investment in environmental protection, the sector was generally poorly addressed.

Following the developments of the 1980s, the UNIP government introduced the National Conservation Strategy in 1985 to enhance guidelines for natural resources management in the country and propose solutions for negative resource exploitation.[9] Although the strategy seemed promising and revolutionary, its overall success was limited by financial constraints that made the implementation of the proposed strategies challenging. In 1990, the government enacted the Environmental Protection and Pollution Control Act (EPPCA), a natural resource protection Act that sought to address the recommendations of the earlier enacted NCS. Although it was not specific to the mining industry, the Act addressed the contamination of air and water, waste, toxic substances, noise, radiation and natural resource conservation.[10] It was therefore important for the mining industry. However, the end of UNIP rule in 1991 meant that the Movement for Multiparty Democracy (MMD) government that rose to power in that year took on responsibility for implementing the Act.

MMD Rule, Privatisation, and Environmental Management in the Mining Industry After 1991

The EPPCA continued to guide environmental management in the country after the change of government in 1991. By this time, Zambia Consolidated Copper Mines, formed through the merging of RCM and NCCM in 1982, while the state remained almost uninvolved, spearheaded environmental management on the Copperbelt mines. By the time the EPPCA was implemented, ZCCM was already undertaking revegetation and tailing treatment activities on the Copperbelt (Limpitlaw and Woldai 1998). In fact, the Environmental Council of Zambia which was established in 1992 reportedly drew several lessons from ZCCM on environmental management, the mining company having already undertaken Environmental Impact Assessments in the region by that time.[11] The state's limited involvement in environmental management on the Copperbelt meant that mining companies were left to self-regulate in this regard. This self-created challenge limited the state's ability to regulate and monitor the extent of environmental decline in the region.

In addition to enabling the creation of the ECZ (now Zambian Environmental Management Agency – ZEMA), the EPPCA also provided for the establishment of the Ministry of Environment and Natural Resources.[12] Other than this, the Act faced several challenges. Similar to the NCS, however, the Act lacked sufficient funding. The provisions of the EPPCA regarding environmental management procedures were also problematic, as they did not specify necessary steps towards their achievement. A major flaw in the EPPCA was its provisions for fines for the emission of hazardous substances. ZCCM was required to pay fines to ECZ for exceeding stipulated emission standards. The fines were introduced to prevent pollution, and as payment towards rehabilitation (Fraser and Lungu 2007: 15–16). However, the fines were minimal and therefore inadequate for environmental mitigation and rehabilitation and did little to prevent pollution on the mines. The proposed compensation system for environmental crimes was therefore not only flawed but also difficult to implement given the self-regulation of ZCCM mining companies at that time.

The (MMD) government decided to privatise the economy, including the mining industry. As was the case during nationalisation, the process was facilitated by legislation – specifically the Mines and Minerals Act of 1995 (Fraser and Lungu 2007: 3). The government claimed that economic reform and improved environmental management of the mining industry were the major aims of privatising the industry (World Bank 2016: 2–3). However, the privatisation of the Zambian mining industry was contradictory. The Zambian Government was faced with a dilemma to either attract the necessary economic development and revenue that was necessary for the mining industry and the growth of the country's economy; or to preserve the environment by providing strict environmental regulation for foreign investors in the Development Agreements signed between the two parties. The environmental provisions were complicated by historic environmental problems that dated back several decades. Ultimately, the Government opted for economic development over environmental sustainability, selling the mines to foreign investors and taking on responsibility for environmental liabilities that it had already failed to address by then (Lungu 2008a: 403–415: Fraser and Lungu 2007: 15–16). This was an enormous responsibility for the state to take on given the financial constraints at the time. This uninformed decision likely resulted from the state's limited involvement in mining-related environmental issues prior to privatisation. This was evidenced by Government haste to engage ZCCM during the negotiations as a way of collecting information on environmental pollution, and the subsequent disagreements between the two parties regarding the accuracy of the environmental status of the mines.[13] As would be expected, the state was less informed about the environmental status of the mining region, given its limited prior involvement in environmental management.

The immediate impact of the terms of privatisation was that mining companies were yet again left without proper environmental regulation, owing to several exemptions granted to mining investors. The consequences were even worse for mining communities on the Copperbelt, which were omitted from the privatisation packages. Communities, therefore, lost access to social service provision, including environmental services such as community gardens that had helped partially address the impact of pollution. Soon thereafter, parks and gardens that were initiated by ZCCM in the communities became dilapidated or almost completely disappeared from townships.[14] Within the industry, the sale of the Copperbelt mines to different investors also created challenges for environmental control, given variations in mining companies and their approaches to environmental management and reporting. In many ways, therefore, privatisation had a negative impact on environmental management on the Copperbelt.

Mining, Pollution, and Environmental Regulation on the Zambian Copperbelt Since Privatisation

An international shift towards sustainable investment occurred around the time privatisation was completed in the early 2000s. Developments such as the 2003 Equator Principles guided large-scale investment in newly privatised countries like Zambia, provided a risk management framework for the environmental and social risks involved in large-scale investment, as well as encouraged environmental reporting (Conley and Williams 2011: 542–575). The 2006 UN Principles reiterated the need for responsible investment and provided guidelines for its achievement (Gond and Piani 2013: 64–104). This was significant for developing countries that relied on foreign investment.

The MMD government eventually sought to reverse some of the conditions of privatisation through legislative reform. The Mines and Minerals Development Act of 2008 repealed the 1995 Mines and Minerals Act that facilitated privatisation. For example, the Act reversed the permission granted to the Minister of Mines and Minerals Development to enter into agreements with investors. Regarding environmental management, the Act stipulated that a potential developer submit an environmental protection plan before development. Furthermore, it granted limitations to the amount of land that could be held by a developer, emphasising that no area larger than that required for the proposed development was to be granted.[15] These conditions jointly reversed some of the previous provisions granted to investors. The revisions sought to improve the state's position. This was partly achieved through the dissolution of clauses that previously bound the Government in the prior Act and Development Agreements, while those concerning mining companies were left in place. The 2008 Act also introduced environmental procedures for mining companies that had not been implemented during privatisation. In this way, the MMD government sought to rectify the problems in mining investment and environmental management created during privatisation.

Despite the strides made at reversing some of the terms of privatisation, certain aspects of the Development Agreements remained in place. Among them were the exemptions granted to investor companies from environmental liabilities. This somewhat limited the achievements of the 2008 Act because mining companies were still within their stability periods and could escape prosecution for emissions not exceeding those previously emitted under ZCCM. Therefore, the provisions of the 2008 Act remained almost ineffective, as most of the stability periods of between 15 and 20 years granted to mining companies had not elapsed by that time (Lungu 2008b: 551).

The Environmental Management Act, that followed in 2011, addressed environmental management, pollution, and waste disposal, as well as sustainable natural resource use.[16] The Act also transformed ECZ into ZEMA. The agency contributes to environmental protection in the mining industries by enforcing environmental regulations and monitoring the activities of all mining operations. Furthermore, it provides guidance to the Government on the formulation of environmental policies. Although the agency has made tremendous strides in regulating mining investment and activities in the country, challenges in securing sufficient finances and work force often limit its effectiveness.[17]

The regulation of the Zambian Copperbelt also draws from legislation that does not directly relate to environmental protection within the mining industry. For example, in recent years, subsidence on the Copperbelt has been addressed through the Disaster Management Act of 2010, which provides strategies for the management of disaster situations. Through the provisions of the Act, Mufulira residents whose homes collapsed from extreme mine-induced cracking previously received tents for temporary shelter.[18] Although this provided a temporary solution, no permanent solution has since been offered to affected residents as even more houses in the area remain on the verge of collapse. Additional laws not specifically relating to mining that are significant to the industry include those addressing water and wildlife management. Among them are the Water Resources Management Act of (Number 21 of 2011) and the Wildlife Acts of 2017.[19] The provisions of the water management law are important where mining activities affect public waterways. The Wildlife Act contributes to the regulation of the mining industry by providing guidelines for the establishment and management of National Parks and fauna, particularly where mining activities threaten the sustainability of wildlife.[20]

Five months after the enactment of the 2011 EMA, the Patriotic Front party won elections, ousting the MMD from power (Rakner 2012: 1). By this time, the Zambian mines were under private ownership, a structure that the newly elected party maintained. The provisions of the Development Agreements, therefore, remained in place. The continuity in the ownership structure meant that the environmental regulation of the industry remained unchanged. Although the Patriotic Front Government went on to amend several existing mining and environmental policies. ZEMA and the Mines Safety Department continued to perform their regulatory functions as before.

Under Patriotic Front rule, enacted laws included the Mineral Resources Development Policy of 2013, Mines and Minerals Development Act (Number 11 of 2015), and the Minerals (General) Regulations of 2016 (Kambani and Masase 2019: 1206–1208). They continue to govern mining activities in Zambia. The Mines and Minerals Development Act of 2015 as read with the Mines and Minerals Development (Amendment) Act of 2018 provides guidelines for the exploitation of mineral resources and the environmental safety of mining investments.[21]

As demonstrated by the various pieces of legislation responsible for regulating the mining industry, no comprehensive laws exist to govern environmental management in the industry. This creates opportunities for both the Government and mining companies to abuse and evade existing laws given the absence of clear guidelines. Furthermore, existing laws contain loopholes that neutralise the strictness of the law (Osei-Hwedie 1996: 61). The strictness of some existing laws has, however, also been condemned as it potentially encourages alternative ways of going around legislative provisions. For example, the 2015 Mines and Minerals Act encourages investors to engage in environmental planning and public engagement to ensure public involvement in decisions regarding the development of new or additional projects.[22] Despite this, the Act lacks a guiding policy framework and does not suggest directives on the inclusion of the public in the licensing process. Due to the absence of guidelines, consultation meetings if held are conducted outside the affected communities, strategically promoting poor attendance. The resulting decisions, therefore, lack input from those directly affected. Despite these misgivings, mining companies that hold consultations in this way would still have met the strict provision of organising a public consultation prior to the proposed development.

100 Years Later: The Copperbelt Today

Today, almost one century since the onset of commercial mining in Zambia, the Copperbelt is a true reflection of the effects of poor environmental regulation in the extractive sector. The political and economic motivation behind the decisions taken regarding its exploitation and regulation have had dire consequences for those for whom the presence of the mines matters the most – local communities, particularly those located closest to mine facilities. Despite the election of a new political party, the United Party for National Development, into power in August 2021, not much has changed in terms of the environmental regulation of the Copperbelt – this partially attributable to the relatively short time spent in office so far. Former mine compounds are a shell of what they were in their heyday. Although a fraction of the residents are currently employed by the mine and can therefore afford a "decent" living, one cannot help but notice the great sense of destitution that fills the atmosphere. Subsidence has become a common sight in many mining towns. What were once flourishing community parks are now abandoned pieces of barren land void of any hope of nourishment for the few animals that seem to roam the community aimlessly. Of these, goats are the most preferred, owing to their ability to consume almost anything. The sight of poorly managed sanitation only adds to the misery, further demonstrating the disastrous fate of the once-flourishing Copperbelt. Several kilometres from here but within the same town, in the effluent areas housing those fortunate enough to live far from the immediate effects of mining pollution, is what appears to be an almost different world, for even though they too are familiar with the stench of sulphur dioxide and feel the occasional tremor, the effects are not as severe for them as they are for those residing in Wusakile, Kankoyo, or Butondo, or at least not yet. For this reason, 100 years later, the effects of mining-induced pollution almost appear to be a myth for those not directly affected by large-scale mining investment.

That some of the provisions of the privatisation Development Agreements remain in place despite legislative reform is disadvantageous for mining communities located closest to the mines that are most affected by pollution, and to environmental regulation in general. Given the omission of mining compounds from the packages offered to investors, mining companies are not legally bound to rehabilitate these communities despite their proximity to the mine and the fact that ongoing mining activities and pollution enhance existing historical environmental concerns. For example, sulphur dioxide pollution has continued to contribute to the destruction of vegetation in mining townships on the Copperbelt and remains a major respiratory health concern. In Mufulira, one of the Copperbelt towns most affected by air pollution through sulphur dioxide emissions, Mopani constructed a sulphuric acid plant in 2014 that reportedly captures 97% of the sulphuric acid previously emitted into the air. According to the mine's Environmental Department, this has significantly reduced sulphur dioxide fumes in the community.[23] The observation contrasts local reports that pollution has in fact worsened and has a more severe effect than before, as residents are now unable to detect the emissions until the effects of exposure to the emissions are felt.[24]

Although complaints may be issued to mining companies following the observation of excessive pollution, it is sometimes difficult to differentiate between historic and ongoing pollution depending on the nature of pollution that occurs. Furthermore, mining companies can evade responsibility for environmental damage on the basis of the exemptions provided during privatisation. For example, mining-induced tremors from ongoing mining have continued to advance the destruction of nearby houses in some mining towns. Despite this, the environmental exemptions contained in the Development Agreements, as well as the challenges in distinguishing the extent of historic versus current environmental damage creates difficulties in holding the mining companies accountable (World Bank 2016: 3). The mine's involvement in the wellbeing of neighbouring communities is in fact limited to Corporate Social Responsibility (CSR), few of which address environmental protection or align with community needs.[25] Hilson raises a similar argument regarding the unsuitability of CSR strategies to the needs of targeted communities (Hilson 2012: 131–137).

Conclusion

As demonstrated in this chapter, the negative impact of mining on the environment, as well as on the health and safety of mineworkers has been significant on the Zambian Copperbelt since the introduction of commercial mining activities almost one century ago. Although knowledge of the environmental impact of mining has increased significantly since the 1920s, this has marked an 'increase' in knowledge, rather than 'emergence' of knowledge about the impact of mining. Despite increased knowledge about mining-related environmental concerns, efforts to address them have generally remained limited. Initial efforts focused on the health and safety of mineworkers, limiting attention to the impact of mining on the wider environment. Although the approach gradually changed, placing greater emphasis on the impact of mining on the wider environment, pollution on the Copperbelt remained rampant. The historic pollution in the region has not been completely addressed to date.

Legislation regarding mining investment and environmental regulation is not effective enough to limit the impact of mining on affected communities. As this chapter reveals, these challenges are closely aligned with local economic and political strategies that have favoured state control as well as increased national revenue and investment at the expense of environmental safety. Furthermore, the 'politics' of state-mining company interactions demonstrate the challenges faced by mineral-rich developing countries, particularly in Africa, in attracting responsible investors without compromising significant areas such as environmental safety. Now, after almost a century of mining and decades of opportunities to improve environmental regulation, the sustainability of one of the world's most significant mining regions is threatened owing to poor environmental management.

Notes

1 Although the term 'Copperbelt' may also refer to the 'new' Copperbelt in the North-Western Province, in this chapter it refers to the traditional 'old' Copperbelt Province.
2 For more information on the severity of tailings disposal see ZCCM-IH, *Horizon*, "The Problem of Tailing disposal", October 1959, pp. 9.
3 For more information on the establishment of the lake see ZCCM-IH, *Horizon*, "Three-Mile long Lake for Mufulira", October 1959, p. 4–7.
4 For additional information on health and safety of workers in the uranium and open-pit mining, see NAZ ML8-1-15, International Mining Conferences and Meetings, Letter from the Commissioner for Mines to the Member for Mines and Works, 8 October 1956.
5 For discussions on Sulphur production in Northern Rhodesia, see variously NAZ ML8-11-83, Mineral Production Sulphur.
6 The examples of racial disparities in disease compensation, see NAZ ML8-1-3, Pneumoconiosis Medical and Research Bureau.
7 For detailed provisions of the laws, see ZCCM-IH, 3.5.3D, Guide to the Mining Regulations, 1973.
8 For more information, see Government of the Republic of Zambia, National Conservation Strategy, 1985.
9 See Government of the Republic of Zambia, National Conservation Strategy, 1985.
10 See Government of the Republic of Zambia, The Environmental Protection and Pollution Control Act, Chapter 204 of the Laws of Zambia, 1990.
11 Interview with Misenge Environmental and Technical Services employee. 2018. Kalulushi. November 6, 2018.
12 See GRZ, The Environmental Protection and Pollution Control Act.
13 For more information of discussions between the state and ZCCM regarding the environmental status of the Copperbelt at the time of privatisation, see ZCCM-IH, 4.2.3J, Privatisation File July 1994 to March 1996, Comments on Strategic Options for the Privatisation of ZCCM Limited, p. 9.
14 ZCCM-IH, 19.4.9E, Mine Township Maintenance and Development (Projects), 1992, Report on Mine Townships Maintenance and Development, 1992, p. 1.
15 See Government of the Republic of Zambia, Mines and Minerals Development Act (Number 7 of 2008).
16 See Government of the Republic of Zambia, The Environmental Management Act (Number 12 of 2011, 87).
17 Interview with Zambia Environmental Management Agency representative. 2017. Solwezi. August 12, 2017.
18 Interview with Kankoyo resident. 2018. Mufulira. November 2, 2018.
19 See Government of the Republic of Zambia, Mines and Minerals Development (Amendment) Act of 2018, Government of the Republic of Zambia, Water Resources Management Act of (Number 21 of 2011) and Government of the Republic of Zambia, Wildlife Protection and Management (Amendment) Act 2017 (Number 5 of 2017).
20 Interview with Trident Foundation Wildlife and Conservation Manager. 2017. Kalumbila, 16 October 2017.

21 See Government of the Republic of Zambia, Mineral Resources Development Policy, (2013); Mines and Minerals Development Act (Number 11 of 2015); Minerals (General Regulations) Statutory Instrument Number 7 of 2016 and Mines and Minerals Development (Amendment) Act of 2018.
22 See Government of the Republic of Zambia, Mines and Minerals Development Act (Number 11 of 2015).
23 Interview with Mopani Copper Mines Environmental Department employee. 2018. Mufulira, 8 November 2018.
24 Variously, interviews with residents of Kankoyo and Butondo, 2018. Mufulira, 2–10 November 2018.
25 The Zambian legislation does not provide clear guidelines for CSR activities within the mining industry. Mining companies therefore outline their own strategies.

References

National Archives of Zambia:
NAZ ML8-1-3, Pneumoconiosis Medical and Research Bureau, 1949–1957.
NAZ ML8-1-15, International Mining Conferences and Meetings, Letter from the Commissioner for Mines to the Member for Mines and Works, 8 October 1956.
NAZ ML8-11-83, Mineral Production Sulphur, 1951–1952.
NAZ ML8-11-99, Ministry of Lands and Mines, Accidents and Diseases, Dust Inhalation, Extract from Hansard Number 85, 10 August 1955.
Zambia Consolidated Copper Mines Investments Holdings Archives Horizon, 1959.
ZCCM-IH, 3.5.3D, Guide to the Mining Regulations, 1973.
ZCCM-IH, 4.2.3J, Privatisation File July 1994 to March 1996, Comments on Strategic Options for the Privatisation of ZCCM Limited.
ZCCM-IH, 19.4.9E, Mine Township Maintenance and Development (Projects), 1992, Report on Mine Townships Maintenance and Development, 1992.
ZCCM-IH, 22.8.7A, Government of the Republic of Zambia. 1971. The Mufulira Mine Disaster Final Report on the Causes and Circumstances of the Disaster which occurred at the Mufulira Mine on 25th September 1970. Lusaka: Government Printers.
ZCCM-IH, C. B. Muchimba, *The Zambian Mining Industry: A Status Report Ten Years After Privatization*, Friedrich Ebert Stiftung, 2010.

Secondary Sources

Adam, C. and A. M. Simpasa. 2011. Copper mining in Zambia: From collapse to recovery. In *Plundered Nations? Successes and Failures in Natural Resource Extraction*, ed. P. Collier, A. J. Venables and T. Venables, 304–348. London: Palgrave Macmillan.
Bebbington, A., A. G. Abdulai, D. H. Bebbington et al., eds. 2018. *Governing Extractive Industries: Politics, Histories, Ideas*. London: Oxford University Press.
Blaikie P. and Brookfield, H., eds. 2015. *Land Degradation and Society*. London: Routledge.
Burdette, M. M. 1984. The mines, class power and foreign policy in Zambia. *Journal of Southern African Studies* 10:198–218.
Conley, J. M. and C. A. Williams. 2011. Global banks as global sustainability regulators? The equator principles. *Law & Policy* 33:542–575.

Dumett, R. 1985. Africa's strategic minerals during the Second World War. *The Journal of African History* 26:381–408.
Gond, J. and V. Piani. 2013. Enabling institutional investors' collective action: The role of the principles for responsible investment initiative. *Business & Society* 52:64–104.
Gray, A. and R. J. Parker. 1929. *The Copper Deposits of Northern Rhodesia*. np.
Hilson, G. 2012. Corporate social responsibility in the extractive industries: Experiences from developing countries. *Resources Policy* 37:131–137.
Jackson, G. C. A. 1933. Outline of the geological history of the N'Changa district, northern Rhodesia. *Geological Magazine* 70:49–57.
Kambani, S. and P. M. Masase. 2019. Mineral rights and sustainable development in the copper mining industry of Zambia: A case study of Lumwana and Kansanshi mines. *Social Science and Humanities Journal* 1195–1210.
Kanu, I. A. 2014. Kenneth Kaunda and the quest for an African humanist philosophy. *International Journal of Scientific Research* 3:375–377.
Larmer, M. 2016. *Rethinking African Politics: A History of Opposition in Zambia*. London: Routledge.
Lungu, J. 2008a. Copper mining agreements in Zambia: Renegotiation or law reform? *Review of African Political Economy* 35:403–415.
Lungu, J. 2008b. Socio-economic change and natural resource exploitation: A case study of the Zambian copper mining industry. *Development Southern Africa* 25:543–560.
Martinez-Alier, J. 2001. Mining conflicts, environmental justice, and valuation. *Journal of Hazardous Materials* 86:214–241.
Monjezi, M., K. Shahriar, H. Dehghani et al. 2009. Environmental impact assessment of open pit mining in Iran. *Environmental Geology* 58:205–216.
Mwaanga, P., M. Silondwa, G. Kasali et al. 2019. Preliminary review of mine air pollution in Zambia. *Heliyon* 5:1–10.
Ndulo, M. 1986. Mining legislation and mineral development in Zambia. *Cornell International Law Journal* 19:1–34.
Osei-Hwedie, B. Z. 1996. Environmental protection and economic development in Zambia. *Journal of Social Development in Africa* 1:57–72.
Paul, R. 1961. Silicosis in northern Rhodesia copper miners. *Archives of Environmental Health: An International Journal* 2:96–109.
Ross, C. 2017. *Ecology and Power in the Age of Empire: Europe and the Transformation of the Tropical World*. Oxford: Oxford University Press.
Saasa, O. 1985. Determinants of Zambia's policies towards the mining industry. In *Issues in Zambian Development*, ed. K. Osei-Hwedie and M. Ndulo, 352–372. Omenana: Roxbury.
Schumaker, L. 2008. Slimes and death-dealing dambos: Water, industry and the garden city on Zambia's Copperbelt. *Journal of Southern African Studies* 34(4):823–840.
Shaw, T. M. 1976a. The foreign policy of Zambia: Ideology and interests. *The Journal of Modern African Studies* 14:79–105.
Shaw, T. M. 1976b. Zambia: Dependence and underdevelopment. *Canadian Journal of African Studies* 10:3–22.
Sikamo, J., A. Mwanza and C. Mweemba. 2016. Copper mining in Zambia: History and future. *Journal of the Southern African Institute of Mining and Metallurgy* 116:491–496.

Sill, K. 2007. The macroeconomics of oil shocks. *Federal Reserve Bank of Philadelphia, Business Review* 1:21–31.
Simpasa, A., D. Hailu, S. Levine et al. 2013. *Capturing Mineral Revenues in Zambia: Past Trends and Future Prospects.* New York: UNDP.
Stern, A. C. 1982. History of air pollution legislation in the United States. *Journal of the Air Pollution Control Association* 32:44–61.
Tembo, A. 2019. After the deluge: Appraising the 1970 Mufulira mine disaster in Zambia. *Historia* 64:109–131.
Vutukuri, V. S. and R. N. Singh. 1995. Mine inundation: Case histories. *Mine Water and the Environment* 14:107–130.

Unpublished Sources and Reports

Chambers, R. and C. R. Gordon. 1991. Sustainable Rural Livelihoods: Practical Concepts for the 21st Century. *IDS Discussion Paper* 296.
Demers, A. 2015. *Bloody Ivory: The Story of Illegal Poaching and Its Global Influence.* Presented to the 17th Annual Africana Studies Student Research Conference held in Bowling Green.
Fraser, A. and J. Lungu. 2007. *For Whom the Windfalls? Winners and Losers in the Privatisation of Zambia's Copper Mines.* Civil Society Trade Network of Zambia Report.
Gibson, C. C. 1994. *Defying a Dictator: Wildlife Policy in Zambia's Second Republic, 1972–1982.* Presented to the Workshop in Political Theory and Policy Analysis held in Bloomington.
Hansungule, M., P. Feeney and R. H. Palmer. 1998. *Report on Land Tenure Insecurity on the Zambian Copperbelt.* Lusaka: Oxfam.
Krueger, J. 2002. *The Basel Convention and the International Trade in Hazardous Wastes.* Yearbook of International Co-operation on Environment and Development.
Limpitlaw, D. and T. Woldai. 1998. *Environmental Impact Assessment of Mining by Integration of Remotely Sensed Data with Historical Data.* Proceedings of the Second International Symposium on Mine Environmental Engineering held in West London.
Rakner, L. 2012. *Foreign Aid and Democratic Consolidation in Zambia.* No. 2012/16. WIDER Working Paper.
Styve, M. D. 2013. *Who Benefits? Norwegian Investments in the Zambian Mining Industry.* Research Report for Caritas Norway.
United Nations Secretary General. 1987. *World Commission on Environment and Development,* Report of the World Commission on Environment and Development.
World Bank. 2016. *International Development Project Appraisal Document on a Proposed Credit on the Amount of US$65.6 Million to the Republic of Zambia.* Report Number 1513.
World Meteorological Organization. 1973. *Special Environmental Report No. 3: Observation and Measurement of Atmospheric Pollution.* Presented to the Technical Conference on the Observation and Measurement of Atmospheric Pollution (TECOMAP), World Meteorological Organization and the World Health Organization held in Helsinki.

6 "Aesthetics of the Earth"[1]
African Literature as a Witness to Post-colonial Ecology

James Ogude

Introduction

There is no doubt that the tendency to see Europe and North America as the epistemological centre of scholarship on ecocriticism and environmentalism is now shifting. Over the last decade or so, there has been a growing awareness that a comprehensive understanding of ecological and environmental discourses is not possible without the inclusion of the rest of the world and especially the global South. The need to move away from what Martinez-Alier (2002) calls the "environmentalism of the poor" produced by privileged subjects for the poor in the South, is now far more pressing than many may have thought (see Guha and Martinez-Alier 1998). We now understand that a genealogy of ecocriticism that overlooks the history of the empire and indeed related colonial inequities may well end up neglecting a long history of critique offered not simply in indigenous narratives of the colonised subjects, but also in post-colonial literatures whose thematic thrust has been nothing but a witness to ecological disaster associated with colonial conquest. Indeed, the argument that the post-colonial field is inherently anthropocentric (human-centred), especially on the African continent, overlooks a long history of ecological concerns in post-colonial literatures. This literature has consistently shown that the process of environmental globalisation is closely tied in with, if not altogether, overdetermined by a long and complex history of the empire. We have to accept that colonial and environmental histories are mutually constitutive because of the central role that exploitation of resources has played in any imperial project, often appropriating these local resources exclusively for transnational corporations.

Elizabeth DeLoughrey and George B. Handley (2011: 3) have, in a compelling argument, reminded us that central to the narrative of exploitation of resources in the history of the Empire is land – lost land that needs to be recovered and re-imagined. It is for this reason that they have commented, with Said in mind "that imagination was vital to liberating land from the restrictions of colonialism and, we might add, from neo-colonial forms of globalization." They make important connections to Edward Said's argument in *Culture and Imperialism* (1994), which seeks to underscore post-colonial literature's role in bearing witness to colonial plunder of land in the colonies and by extension the broader ecologies, while simultaneously re-imagining its recovery. Said, as DeLoughrey and Handley point out, was attempting to surface the ecological function of post-colonial literature, a phenomenon that we would begin to witness more fervently in African literature of the 1960s and beyond. In the 1960s and the following decades, land became the key ecological trope in this project of restoration, especially in contesting colonial notions of virgin nature – untamed wilderness that is in need of conquest and taming – a dominant trope that one comes across even in colonial settler literature in Africa. It is for this reason that Simon Gikandi (1984: 234) has written: "The image of the settler mapping new frontiers and taming wild Africa to his will is a common one in colonial writing." He adds: "Through the whole breadth of the Eastern African region covered in [Blixen's] *Out of Africa*, for example, what is always underscored is the perspective of the aroused imagination, of the nobility of the land and its wideness" (234).

Thus, it is correct to argue like DeLoughrey and Handley (2011: 12) that

> Colonial violence was mystified by invoking a model of conserving an untouched (and often feminized) Edenic landscape. Thus the nostalgia for a lost Eden, an idealized space outside of human time, is closely connected to displacing the ways that colonial violence disrupted human ecologies.

Therefore, when the African writers begin to write, especially in settler colonies, their primary task as Frantz Fanon (2004) intimated, is to talk about the reconstitution of land as a source of life and as an extension of the human. It is also an attempt to inscribe post-colonial ecology as a history of displacement, dispossession, and a degradation of nature itself. One vital aspect of post-colonial ecology was to reimagine this displacement between people and place, while equally displaying a longing for ethics of place and belonging – a certain element of human continuity and affinity with the land, which has been the defining feature of foundational narratives in Africa (see Achebe 1958; Plaatje, 1978).

In response to the history of the empire, especially here in Africa, Ngugi has been adept at revising this colonial image of land in Africa. In what follows, I want to turn to a range of related issues that have informed his engagement with ecology and environmentalism. Firstly, I want to demonstrate how some of his texts, in particular, *Weep Not Child* and *Petals of Blood*, have focused on land as a witness to environmental degradation by insisting that we cannot separate the history of the empire from ecocritical thought because in doing so we de-historicise nature. The intention here is not to create rigid geographical differences, but rather to point to how a close reading of African literature provides a rich and nuanced discussion of ecological degradation in the wake of the colonial experience. Secondly, I seek to show how Ngugi offers a critique of modernity, especially in the way its consumerist ideology and waste compound problems of environmental degradation, thus drawing attention to the economics of human ecology as a vital historical aspect of post-colonialism. Finally, I ask whether Ngugi's works, in spite of their awareness of ecological and environmental challenges, inscribe poetics of regeneration and aesthetics of the earth – of renewal and possibilities for a better environment.

If there is one lingering aspect of African literature's response to the effects of colonial modernity, it is in the way it firmly places the human in nature. There is no better example of this other than in the works of the Kenyan writer, Ngugi wa Thiong'o whose works grapple with the structural impact on the Kenyan societies by colonialism. These include land alienation and other related imposition of values and discursive practices whose ultimate effect was the dislocation and displacement of Africans from their material and spiritual being. The impact of colonial modernity appears in most of Ngugi's works such as *Secret Lives, The River Between, Weep not Child*, but perhaps most eloquently dramatised in his foremost proletariat novel, *Petals of Blood*.

In *Secret Lives*, the story "Gone with the Drought" captures one of the most moving images of disaster – a landscape scorched by the sun:

> From ridge up to ridge the neat little shambas stood bare. The once short and beautiful hedges – the product of land consolidation and the pride of farmers in our district, were dry and powdered with dust. Even the old mugumo tree that stood just below our village, and which was never dry, lost its leaves and its greenness – the living greenness that had always scorned short-lived droughts. Many people had forecast doom. Weather prophets and medicine men – for some still remain in our village though with diminished power – were consulted by few people and all forecast doom.
>
> (wa Thiong'o 1975: 15–16)

The ecological disaster captured in this short passage is best framed in the words of the narrator who tells us that the landscape appears to have been gripped by "the whiteness of death" (1975: 2). The deliberate use of "whiteness" here is not lost on us for we are not only talking here of the impact of colonial land alienation figured in the patches of land consolidation, now belonging to African farmers who are totally oblivious of the impending doom that they have been squeezed into, but perhaps more importantly, the concomitant cultural landscape that has been decimated by colonial conquest and the subsequent loss of locality to the outsider. The African farmers in this story have been forced into a new ontological relationship with the land, not simply through the act of land alienation, but equally through a colonial structure that now restricts the farmers to consolidated pieces of land that must of necessity be overworked over time. No wonder then that the resultant effect is the drought that does not spare even the mugumo tree that would seem to have been the constant symbol of ecological stability in the village. The weather prophets of the village have also been consigned to the margins of society in a context where "the whiteness of death" reigns.

96   James Ogude

Ngugi seems to be suggesting here that in order to grasp a historical model of ecology and an epistemology of space and time in a post-colonial context, we have to enter into a deep historical dialogue with the landscape itself; to capture the trauma of conquest, even when on the surface of it an idealised modernist idea of landscape is offered.

In *Weep Not, Child*, Ngugi sharply introduces us to the ontological value of land among his people when the narrative voice talks about Ngotho and his reaction to land alienation:

> for to him it was a spiritual loss. When a man was severed from the land of his ancestors where would he sacrifice to the Creator? How could he come into contact with the founder of the tribe, Gikuyu and Mumbi? What did Boro know of oaths, of ancient rites, of the spirits of the ancestors?
>
> (1964: 74)

The spiritual significance of the land explains the reverential awe with which Ngotho treats it. It is not surprising that although Ngotho at one point thinks that education is everything and urges his son, Njoroge, to go to school, he is doubtful about the efficacy of education "because he knew deep inside his heart that land was everything. Education was good only because it would lead to the recovery of the lost lands" (Ibid.: 39).

Ironically, it's their feeling for the land that holds Ngotho and Howlands, the colonial settler, incongruously together: "Not that Mr Howlands stopped to analyse his feelings towards him. He just loved to see Ngotho working on the farm; the way the old man touches the soil" (Ibid.: 29–30). Thus what superficially unites them, at a deeper level divides them irrevocably. Ngotho is at one with land, co-operates with it; Howland triumphs over it, possesses it assertively as a conqueror:

> They went from place to place, a white man and a black man. Now and then they would stop here and there, examine a luxuriant green tea plant, or pull out a weed. Both men admired this *shamba*. For Ngotho felt responsible for whatever happened to this land. He owed it to the dead, the living and the unborn of this line, to keep guard over this *shamba*. Mr Howlands always felt a certain amount of victory whenever he walked through it all. He alone was responsible for taming this unoccupied wildness. They came to a raised piece of ground and stopped. The land sloped gently to rise again into the next ridge and the next. Beyond Ngotho could see the African Reserve.
>
> (*Ibid.*: 31)

Whereas for Mr Howlands, the adventure presents itself as a form of triumph and domination over nature, for Ngotho it is an ontological rupture and anguish at the loss of land. The loss is far much deeper than a material loss; it is a total rupture, which signals an impending ecological disaster that only history can allow us to understand. This rupture and displacement is captured in a sharply divided landscape between the vast white settler land and the African Reserves that mark the beginnings of a new relationship with the environment – a relationship that now defines the erstwhile owners of the land – the Africans – as squatters on their own environment.

Although Ngugi's account here may come through as romantic and suspended on stark binaries between the white man and the black man, and in particular their relationship with land, it is nevertheless important in drawing our attention to the sense of pathos that marked those initial moments of displacement of the Africans from their land – the totality of their environment. It marked the beginning of what Alfred Crosby (2004), calls "ecological imperialism" – a phenomenon whose effects were felt in different ways throughout colonial Africa. It also signalled the introduction of the Cartesian logic in an African environment, thereby destabilising a long abiding relationship between humanity and nature and that is the pre-colonial metaphysical social genesis of the individual and his/her dependence not only on others but on the totality of its environment. It is what Cajetan Iheka calls "an ethics of the earth" found in many African societies. "In this mode of seeing," Iheka adds, "certain nonhuman forms, including animals, plants, and so on, are considered viable life worthy of respect" (2018: 7). It also gestures towards the post-humanist turn on ecocriticism (post ecocriticism) which is a more engaged mode of the symbiotic and co-evolutionary relationship between organic and inorganic forms of life, suggesting the need for sustainability and acting responsibly towards our planetary universe.

Significantly, Ngugi extends and complicates the idea of ecological imperialism, introduced in his earlier texts, in *Petals of Blood*; a text that is a vivid dramatisation of the brutal force of colonialism and its transformative nature, occasioned by the violent imposition of a new mode of production whose purpose is the exploitation of the human and natural resources of the colonised subjects. It is this phenomenon that the text contests.

My interest in Ngugi's *Petals of Blood* is in the way that it draws attention to the Africans and their natural environment, a theme rarely examined in the novel, and yet it is most apparent in Ngugi's handling of the relationship between the African people and nature, and the harmful exploitation of the land. The centrality of environmentalism is also foregrounded through those motifs that signal resistance to the dark vision imposed by colonial modernity. In this respect, the title of the novel to which I will return shortly signals one such utopian gesture and the capacity for regeneration and restoration of the human in nature. It is an affirmation of what Iheka calls "aesthetics of proximity" – an expression that not only designates occupying the same space but also collapsing of the artificial dichotomy between the humans and non-humans (Iheka 2018: 17). The contestation between Munira, the teacher, and one of his pupils over the colour of a certain flower, which resembles the colour of blood is important in opening up some of the theoretical issues at play here. When Munira hears from his pupil that one of the flowers resembles the colour of blood, his response throws a range of issues into sharp relief. His response is telling:

> There is no colour called blood. What you mean is that it is red. You see? You must learn the names of the seven colours of the rainbow. Flowers are of different kinds, different colours. Now I want each one of you to pick a flower . . . count the number of petals and pistils and show me its pollen
>
> (1977: 21).

At the heart of the contestation between Munira and his pupil is a Cartesian notion of nature born of Munira's colonialist education on the one hand, and a relatively organic view of the natural world that the student brings to the surface, a view which unlike the teacher's relates to the colour of the flower and his own blood. The relationship does not distinguish between subject and object and is founded on sensory perceptions as opposed to the concept of the plant as an object of analysis, minutely categorised according to the scientific rules of plant anatomy. The student's question disturbs the teacher precisely for extrapolating the limits of such rationality. Munira can be seen as an individual who has been alienated by the Enlightenment thought of colonial modernity, a subject-object that attempts to find in objective knowledge an escape for the fear of what one cannot dominate. As Max Horkemer and Teodor Adorno argued in their 1945 book, *Dialectics of Enlightenment* (2002), the domination of nature translates into the domination of other humans as the history of colonialism shows. For the authors, the attempt to overcome mythical world views through a violent process of rationalisation impoverished and reified men's ability to think and relate to nature, which became a mere object of speculation and experimentation. One of the reasons for this desperate clutch at rationalisation is men's fear of the unknown.

In Munira's case, fear surfaces through the spontaneous manifestation of his student to whom nature is not classifiable according to parameters that are strange to daily human experience. In the process, the man-nature based relation – the alleged sacredness of natural elements and ancestrality is, gradually, transformed in the novel into a relation in which nature becomes an object of study and exploitation with the function of producing material richness to the colonisers and to African elites to the detriment of ordinary Kenyans. A new regime of truth, in Michael Foucault's sense, emerges by which "we are judged, condemned, classified, determined in our undertakings, destined to a certain mode of living or dying, as a function of the true discourses which are the bearers of the specific effects of power" (1980: 93–94).

In *Petals of Blood*, the new regime of truth ensures that nature, previously inseparable from the African person and Kenyan societies' modes of living, must now be apprehended only through the mediation of reason. This mediation has the effect of distancing the subject from the natural world

"Aesthetics of the Earth" 99

and works as a support, in the intellectual plan, to the extractive policy implemented by the British government. That is, for nature and the human to be completely dominated, it is fundamental that Africans think of nature no longer as a coherent whole, but as an object of exploitation. The novel is most forceful in its dramatisation of this when it later shows deforestation and draught as the major consequences of the progress and development brought by colonial modernity. What happens to Ilmorog, the novel's main setting, is a clear witness to the character of ecological and environmental degradation – a devastating desertification that one of the characters of the novel draws our attention to:

> You forget that those days the land was nor for buying. It was for use. It was also plenty, you need not have beaten one yard over and over again. The land was also covered with forests. The trees called rain. They also cast a shadow on the land. But the forest was eaten by the railway. You remember they used to come for wood as far as here – to feed the iron thing. Aah, they only knew how to eat, how to take away everything.
> (1977: 82)

There is clearly a deep understanding of ecological balance and environmental sustenance among these people and yet this indigenous resource base is now threatened by a new regime of truth – a scientifically demythologising discourse. The new regime of knowledge is followed by the establishment of the capitalist mode of production, which creates untold poverty and produces a sharp division between the local bourgeoisie and the newly constituted proletariat – the dispossessed peasants. The representation of the railway as a "tree-eating" monster is a grotesque and graphic rendering of the impact of capital on nature – commodified African landscape, which is now subjected to a rapacious and predatory consumption leading to environmental devastation and deforestation. Linked to this is the commodification of *Theng'eta*, a drink produced from the flower with petals of blood and used by the community in specific ceremonies. Its commodification points to the disruptive nature of the new system, which as Patrick Williams argues, is how "capitalism turns use value . . . into exchange value" (1999: 83). *Theng'eta*, which was, formerly, an important symbol of unity for the community, totally loses such meaning as it is reified by the capitalist mode of production. The spiritual and communal reach of it, as well as the history of its production and use by the people, disappears in the industrial production and in the mass consumption of an unspecified product.

If we return to the metaphor, "petals of blood" for a moment, we are tempted to read two competing versions of meaning: one the scientific rationality imposed by modernity and the other a culturally and socially based reading of the world at war here. Critics such as Laura Wright have observed that the metaphor points to the devastating effects of a capitalist-driven development on land and the violence visited on a post-colonial

Kenya that can only pointlessly return to an elusive past (2011). Wright writes: "the image of the bleeding flower . . . is an image of the corruption and devastation of nature as a result of deforestation and the construction of Trans-Africa highway" (2011: 19–55).

There is no doubt that the epistemic and physical violence that leads to the transformation of the local culture and the deforestation of the region and the complicity of the new local elite in the plunder of their country is a major theme in the novel. There is certainly sinister and slow violence at work, as Rob Nixon (2011) would have it. However, the metaphor of "petals of blood," when properly understood also signals something much more complex than the surface symbol of violence it evokes. It has to be seen as reflecting the power of an uncontrollable nature that somehow resists the enlightenment thought and capitalist reification. When Munira burns down Wanja's brothel, and witnesses "the tongues of flame from the four corners forming petals of blood" (333), he is convinced that he is engaged in a regenerative act; the inauguration of a new beginning. Inspector Godfrey intimates as much but goes further in his acknowledgement that although a certain element of moral purity is desirable for the rotten system to survive, the problem occasioned by the capitalist mode of production is layered, complex, and goes way beyond the surface symbolic structures that we see. The narrative voice observes:

> This system of capitalism and capitalist democracy needed moral purity if it was going to survive. The skeletons that he himself had come across in Nairobi, Mombasa, Malindi, Watamu and other places he had never given much thought . . . And it was not Wanja's Sunshine Lodge that Inspector Godfrey was thinking about. It was, for instance Utamaduni Cultural Tourist Centre at Ilmorog. Ostensibly it was here to entertain Watalii from the USA, Japan, West Germany, and other parts of Europe. *But this only camouflaged other more sinister activities: smuggling of gemstones and ivory plus animal and even human skins. It was a centre for plunder of the country's natural and human assets* (1977: 334).
>
> (My emphasis)

The reality is that leisure and its consumption is inextricably linked to the exploitation of human and natural resources, and the expropriation of land emphasised in the novel. But in spite of the commodification of nature, the exploitation of human resources, and the expropriation of land emphasised here, we are still left in no doubt that the expression "petals of blood" appears, alternatively, as that which one cannot colonise, i.e., that which somehow resists the capitalist mode of production and modernity's logic. From this perspective, even though the novel presents a reality radically transformed by colonial modernity, the title signals to a possible way out or, at least, to an uncertain future founded on the symbolic forces of the present.

This too is what is embodied in Munira's fire which according to Patrick Williams "marks, if not an end of an era, at least a significant moment with transformative potential; to Karega's final vision of the reign of the bloodthirsty monsters which must be swept away to create the opportunity for the flowering of full human existence" (wa Thiong'o 1977: 80). I understand full existence to mean one in which humanity returns to nature, not to dominate it, but to be at one with it – a clear evocation of the principle of co-agency embedded in many African eco-systems.

The regenerative spirit foregrounded here is linked also to a possible redemption that had already been signalled in the specific part of the novel called "The Journey." This is the context in the novel in which men, women, and children undertake an impossible journey through the desert to demand nourishment and political support in Nairobi – the seat of capital – after one year of drought whose cause we now know only too well – the extraction of the rain-bringing forest. Although the journey pushes the community to their limit, it also signals a new beginning of unity and a vision for change; change for a new aesthetics of the earth in which "all flowers in all their different colours would ripen and bear fruits and seeds. And the seeds would be put into the ground and they would once again sprout and flower in the rain and sunshine" (Ibid.: 303).

If the vision sketched above appears utopian, the narrative voice is equally realistic when it asserts that "history and legend showed that Ilmorog had always been threatened by the twin cruelties of unprepared-for vagaries of nature and *the uncontrollable actions of men*" (Ibid: 111) – (my emphasis), thus forcing the reader's to confront the politics of the Anthropocene and for the post-colonial world to regain its agency – to see its complicity in the degradation of the earth and in the exploitation of its resources. This kind of awareness and agency, Ngugi would seem to suggest, is needed in confronting post-colonial ecology without having to reduce everything to the logic of colonial modernity. It is for this reason that the community of Ilmorog rejects a hollow tradition of sacrifice targeted at the outsider's donkey because as Karega reminds the community of Ilmorog:

> A donkey has no influence on the weather. No animal or man can change a law of nature. But people can use the laws of nature. The magic we should be getting is this: the one which will make this land so yield in times of rain that we can keep aside a few grains for when it shines.... That magic is in our hands. Tomorrow when it rains: we should be asking the soil: what food, what offering does it need so that it will yield more?
> (*Ibid*: 114–115)

This balanced eco-system in which nature makes its demands on humanity for sacrifice, for rains and better yields, and humanity in turn yields to its affinity with nature, speaks to the broader ethical and ontological question of earth keeping and its regeneration embedded in a particular African ontology of co-existence and conviviality embodied in the Southern African philosophy of Ubuntu.

Ubuntu is not simply about the awareness of our interdependence as humans, it is also more importantly about the struggle to eschew self-referentiality, domination, and possession – the defining ethos of colonial and post-colonial worlds. Ubuntu at its best is about our planetary conviviality in which humans are only one among other elements that constitute its well-being. It is, in many ways, and in a profound sense, a post-humanist ecological thinking. Ecology is here understood not in terms of a human-centred, but as a radical mapping of the traditional epistemic terrain, which tends to valorise the human in any ecological discourse. The ethics and politics of post-colonial ecology is therefore a new type of radical ecology that shows deep affinity to nature and is equally respectful to all creatures.

In conclusion, although Ngugi's texts discussed here may not be entirely representative of Africa's post-colonial ecology, they nevertheless point to important issues pertinent to the relationship between humanity and nature in an African context. Firstly, texts demonstrate that the domination of nature or land in its broadest sense (sea included), translates into the domination of other humans. The domination of other humans is therefore intertwined with environmental degradation. Secondly, the texts teach us that environmental concerns have always been very much part of ethical value systems among Africans and the colonised subjects in general, and not the exclusive prerogative of the privileged north. Thirdly, the battle for liberation has always been inextricably bound up with the battle for land restoration and against resource extraction and exploitation of the earth. Fourthly, Africans have always problematised environmental issues in ways that place the human in nature. Long before the rise of global discourse rooted in a human environment emerged in the 1970s, African scholars like their Third World counterparts were already grappling with these issues; insisting that there should be no decoupling of humanity from the natural world. African writers like Ngugi had long introduced the economics of human ecology, which for many may have appeared anthropocentric, while in fact what they were drawing attention to was environmental sustainability that rooted the problem of formerly colonised societies within a long history of colonialism and globalisation. Although Ngugi's texts grapple with colonial legacies and realities of the post-colonial experience after it, they also point to the need for the agency; for insurrectionist struggle against rampant exploitation of the earth by the colonial settlers and the black political elite that captured power at the time of independence.

Finally, Ngugi like many of his contemporaries linger on the moment of encounter with white settler colonialism in order to surface those spiritual ties that bind the Africans to the land, it is not to romanticise this relationship or to ground his critique on some primordial natural purity as many have argued, but rather to point to African societies well-known ethic of tendering the earth. It is a negation of the anthropocentric approach to environmentalism and equally to place into stark contrast the damage that was occasioned by colonialism and the extended history of the empire. As DeLoughrey and Handley have noted: "To separate the history of empire from ecocritical thought dehistoricizes nature and often contributes to a discourse of green orientalism" (2011: 20). Besides, discourses of precolonial purity and harmony have to be seen as powerful rhetorical and strategic devices often wielded, in the hands of writers like Ngugi and Saro-Wiwa, among others, to unsettle the self-assuring discourses of rationality suspended on European Enlightenment thought and to embrace what Dipesh Chakrabarty (2000: 239), calls multiple ways of being in his study about "provincializing Europe."

The authority of Ngugi's discourse on post-colonial ecology is that it strives to collapse the hierarchies between the human and nonhuman; humanity and nature dyad. For example, there is a fundamental ethic of care at work that transcends the human desire to dominate nature when we watch Ngotho's relationship with land. This is what Edouard Glissant calls "the aesthetics of the earth" – that capacity to celebrate beauty against the backdrop of loss and ravaged land.

It is only in this sense alone that we can understand the journey that the community of Illmorog undertakes in *Petals of Blood* against the background of drought and hunger. It is an affirmation of "aesthetics of the earth" as a discourse not simply embedded in disruption and connection; disunity and unity; but also a refusal to accept conditions of displacement and fragmentation. It is equally a refusal of wholeness – of unity of the subject that modernity entails – although more importantly, it is a search for new epistemologies, not simply to understand the earth's degradation, but to provide new ways of thinking and new ways of being; new hermeneutics and possibilities for the earth's sustenance that only scholarship rooted in global humanities can offer. It is in this sense that we see Post-humanities and Environmental Humanities as establishing a symbiotic and sustainable nexus – a symbiocene. By symbiocene, we mean a new era after the Anthropocene, defined by a symbiosis that implies living together for mutual benefit and affirming the interconnectedness between human and non-human things; life and all living things – Ubuntu.[2]

Notes

1 I owe my understanding of this term to DeLoughrey and Handley (2011) who evoke this phrase in relation to Edouard Glissant's use of this term in the context of the Caribbean. According to Edouard Glissant, "Aesthetics of the earth" works to affirm beauty even in contexts where land and the sea have been ravaged by colonial violence. He makes the point that the apparent inappropriateness of aesthetics in the context of waste and rupture can enable a regenerative response by channelling people's energies towards "a love of the earth so ridiculously inadequate or frequently the basis for sectarian intolerance" (1997: 15). This way, Glissant hopes that literature can teach the world the political force of ecology by showing "the relational interdependence of all lands, of the whole earth" (1989: 147).

2 For further readings on Ubuntu, see James Ogude (ed.) *Ubuntu and Personhood*. Trenton, NJ: Africa World Press, 2018; *Ubuntu and the Reconstitution of Community*. Bloomington: Indiana University Press, 2019; James Ogude and Uni Dyer (eds.). *Ubuntu and the Everyday*. Trenton, NJ: Africa World Press, 2019).

References

Achebe, C. 1958. *Things Fall Apart*. London: Heinemann.
Adorno, T. and M. Horkheimer. 2002. *Dialectic of Enlightenment: Philosophical Fragments*. ed. G. S. Noerr, trans. Jesophott. Stanford: Stanford University Press.
Blixen, K. 1994. *Out of Africa*. London: Cape.
Chakrabarty, D. 2000. *Provincializing Europe: Postcolonial Thought and Historical Difference*. Princeton: Princeton University Press.
Crosby, A. W. 2004. *Ecological Imperialism: The Biological and Cultural Consequences of Europe 900–1900*. 2nded. Cambridge: Cambridge University Press.
DeLoughrey, E. and B. G. Handley. 2011. *Postcolonial Ecologies: Literatures of the Environment*. New York: Oxford University Press.
Fanon, F. 2004. *The Wretched of the Earth*. Trans. R. Philcox. New York: Grove Press.
Foucault, M. 1980. *Power/Knowledge: Selected Interviews and Other Writings, 1972–1977*. New York: Pantheon Books.
Gikandi, S. 1984. The growth of the east African novel. In *The Writings of East and Central Africa*, ed. G. D. Killam, 231–246. London: Heinemann.
Glissant, E. 1989. *Caribbean Discourse: Selected Essays*. Trans. J. Michael Dash. Charlottesville: University of Virginia.
Glissant, E. 1997. *Poetics of Relations*. Ann Arbor, MI: University of Michigan Press.
Guha, R. and J. Martinez-Alier. 1998. *Varieties of Environmentalism: Essays North and South*. Delhi and New York: Oxford University Press.
Iheka, C. 2018. *Naturalizing Africa: Ecological Violence, Agency, and Postcolonial Resistance in African Literature*. Cambridge: Cambridge University Press.
Martinez-Alier, J. 2002. *The Environmentalism of the Poor*. Cheltenham, UK and Northampton, MA: Edward Edgar.
Nixon, R. 2011. *Slow Violence and the Environment of the Poor*. Cambridge, MA: Harvard University Press.
Ogude, J. 2018. *Ubuntu and Personhood*. Trenton, NJ: Africa World Press.
Ogude, J. 2019. *Ubuntu and the Reconstitution of Community*. Bloomington: Indiana University Press.
Ogude, J. and U. Dyer. 2019. *Ubuntu and the Everyday*. Trenton, NJ: Africa World Press.
Plaatje, S. 1978. *Mhudi* (First Published by Lovedale Press, 1930). London: Heinemann.
Said, E. 1994. *Culture and Imperialism*. London: Vintage.
wa Thiong'o, N. 1964. *Weep Not, Child*. London: Heinemann.
wa Thiong'o, N. 1975. *Secret Lives*. London: Heinemann.
wa Thiong'o, N. 1977. *Petals of Blood*. London: Heinemann.
Williams, P. 1999. *Ngugi wa Thiong'o*. Manchester: Manchester University Press.
Wright, L. 2011. Inventing tradition and colonizing the plants: Ngugi wa Thiong'o's *Petals of Blood* and Zakes Mda's *The Heart of Redness*. In *Environment and the Margins: Literary and Environmental Studies in Africa*, ed. B. Caminero-Santangelo and G. Myers, 19–55. Athens: Ohio University Press.

7 "One in Heart as They Are in Tongue"

"Yoruba," Land and Environmental Violence in Colonial Southwestern Nigeria

Ayodeji Wakil Adegbite

Introduction

In 1888, the first British Governor of Lagos, Alfred Moloney (1886–1891) wrote a letter to the Alaafin of Oyo, in Southwestern Nigeria. In the letter, Moloney stated: "I always aim at making all Yoruba-speaking peoples one in heart as they are in tongue." He continued, "towards such unity I attach importance to a definite and permanent understanding between these Yoruba-speaking peoples, and this colony which is mainly inhabited by Yorubas."[1] The quote from Governor Moloney, from which I draw the topic of this chapter, captures a paradox of British colonialism in Africa that has far-reaching implications than is often examined by scholars of African history. It summed up how, beyond guns and bayonets, colonial authorities relied largely on the formation of ethnic identity as a means to enforce colonial domination on both human and environment in Africa (Culture and Imperialism 1993). The "unity" between "Yoruba's" "heart" and "tongue" that Moloney's letter proclaims, is not only built on a colonial assumption that limits sociality to humans. Indeed, Moloney assumes agency simply for the Yoruba humans inhabiting the region and assumes that interactions are exclusively within these humans, with "nature" seen as a separate constituent of the society and simply meant to be exploited.

Moloney's attempt at strengthening Yoruba kinship, albeit critical to the colonial capitalist expansion and environmental domination, is antithetical to the South-western region's indigenous human–environment kinship. Indeed, British authorities have, since the conquest of Lagos in 1861, gradually shifted the ontological premise of indigenous ecosocial relationship through the use of treaties, territories, and later land tenure systems. Since the British authorities began to encroach into the Southwestern region of what would become Nigeria, it has ignored and made concrete efforts to erase indigenous relationships with land, non-human, and spirits and the reverence that indigenous groups ascribe to them. The "unity" Moloney aspired to was antithetical to the thread that held the indigenous groups in common with the ancestors, other-than-humans, and future generations.

DOI: 10.4324/9781003287933-7

This chapter explores the changing human–environment relationship in colonial Southwestern Nigeria that led to the flourishing of colonial extraction economies and "modern" industries that haunt the present and future of this beautiful planet. The "success" of mineral exploitations and palm oil extraction in 19th century colonial Africa inspired and consolidated colonial remaking of indigenous environmental knowledge. Colonial incursion reorganises knowledge of land management through treaties, ordinances, and tenure systems. The sort of unbridled exploitation of "natural resources" that follows is in opposition to indigenous precolonial human–environment kinship. In this chapter, I argue that the Anthropocene and environmental violence is a historical and colonial process, one that was predicated by the insidious erasure of indigenous human–environment kinship.

Moloney's assumption as presented in his letter has a discernible route. African historians of the environment in different regions of the continent from South to North Africa and East to West Africa provide ample evidence of a European colonialists ideology that impose European image upon the African landscape.[2] This colonialist ideology bares an ontological marker of Western modernity that separates the domain of the "natural" from human society. This ideology stems from the Cartesian worldview that identified the mind as being separate from the corporal body. This ontological premise supports the exploitation of natural resources – since nature is considered a resource for humanity – which accompanied the development of capitalism and the European colonial project. Yet, unlike in Europe, there is no essential dualism between the human society and nature in many African indigenous worldviews. No distinguishable marker separates the mind and the body. In precontact "Yoruba" cosmologies for example, balance is aspired to and achieved through a broad social catalogue that incorporates entities in the society, including spirit, ancestors, land, and other species (Oluwole 1997, 2014).

In *Misreading the African Landscape* Melissa Leach and James Fairhead's using a range of evidence and methods, fleshed out African sustainable agro-ecological knowledge that captures the human–environment kinship I describe in this chapter. In the face of French colonial officials and "experts" (mis)interpretation of landscape in the savannah zone in Guinea, Leach, and Fairhead found that "elders and others living behind the forest walls provide different readings of their landscape" – as filling with forest, and not "emptying of it" and as formed by them and their ancestors (Fairhead and Leach 1996: 2–3). African indigenous knowledge and kinship with trees, ecology, and the "sophistication of local land, soil, vegetation, and management techniques and the wisdom and creativity of indigenous science" shows that the patches of forests is "associated with vagrant, mobile impermanency" and not declining as "experts" and colonialists would have Africans believe (Ibid. 6). Yet, due to colonial misreadings, assumptions, and limited knowledge of African human–environment kinship, colonial powers claim that African societies lacked scientific knowledge and were too racially inferior to understand and manage their environment. Colonialists thereby imposed their ideologies on Africans and the African landscape and consolidated this belief with the creation of treaties, tenure systems, and ordinances. These processes facilitated extraction regimes that strengthened the brutal structures of capitalism which is now firmly rooted in post-colonial Africa and helped created inequitable environmental burdens of the so-called Anthropocene. In this chapter, I showed the complicity of African elites to the erosion of indigenous eco-social relationships.

This chapter provides a fresh perspective to a well-told history of "colonial conquest" from the lens of environmental humanities/history. It explores contemporary environmental crises of the Anthropocene as a historically situated and inequitable process, whose footprint is developed in colonial Africa. The Anthropocene (anthropo meaning "human" and cene meaning "new") is the new age, and the scientific designation of a planetary phase where "humans" have become a geomorphic force that influences the functioning of Earths systems (Crutzen 2002). Yet saying "human" "attributes the ecological collapse to an undifferentiated 'humanity' when in practice both responsibility and vulnerability are unevenly distributed" (Hecht 2018). The term Anthropocene also takes its root from the Western polarisation of nature versus human. The nomenclature, therefore, tends to silence the ideologies of African indigenous societies that are often the hardest hit by the effects of climate change but, whose ideology does not conform with this worldview. Indeed, the Anthropocene label masks histories of colonialism, global commerce, epistemic violence, that built and sustained capitalist regimes that drove economic growth and consumption the world over (Moore et al. 2019). Told anew, African colonial histories can shed light on the hidden geographies, colour line, and insidious markers of the Anthropocene (Yusoff 2018).

Although various start date or golden spikes has been proposed as the beginning of the Anthropocene, including the 1950s atmospheric testing of nuclear weapons. Yet scientific golden spikes often align with media spectacular that hides insidious violence of environmental problems. In *Slow Violence and the Environmentalism of the Poor*, Rob Nixon argued that environmental violence can be insidious so as not to draw spectacular attention. It might be in form of a slow violence "that occurs gradually and out of sight, a violence of delayed destruction that is dispersed across time and space, an attritional violence that is typically not viewed as violence at all" (Nixon 2013). In an era of media "spectaculars," the scenes of environmental violence might be trapped within unequal global commerce and interconnections where African bodies and landscapes are racialised, erased, and made invisible. Therefore, while scientists are looking at nuclear debris or such other spectacular fingerprints of humanity for the geological epoch, Africa provides one such medium of insidious pointer to this geologic time.

Massive palm oil extraction in Africa, for example, in the words of historian Martin Lynn "grease(d) the wheels of the industrial revolution" in the early 19th century. Scientists that pin the golden spike to industrial revolution will do well to account for the role that massive palm oil extraction from Africa played in the Anthropocene (Lynn 2002). Anthropologist, Gabriel Hecht has also investigated the effect of pollution and radioactive deposits from South African gold mining and uranium respectively in the 19th century on Africans and how that contributes to the Anthropocene (Hecht 2018).

Indigenous scholars from both Africa and North America hold a similar view that colonial impact on indigenous ecosocial knowledge constituted a form of environmental violence. Vanessa Watts in the case of the colonisation of Haudenosaunee and Anishnaabe people in the America argues that "the measure of colonial interaction with land has historically been one of violence . . . where land is to be accessed, not learned from or a part of" (Watts 2013; Whyte 2018). This disruption of Indigenous relationships to land, and environment according to Tuck Eve, and Wayne Yang represents a profound epistemic, ontological, and cosmological violence (Wayne and Tuck 2012). The process of environmental violence in Southwestern Nigeria began with the creation of the Yoruba identity which was first inspired by Yoruba elites and later taken up by colonialists. What follows is the categorisation of land and palm produce as commodities and as entities without agency, disconnected from human.

In Africa today, there is a global rush for land with estimates of millions of acres sold or leased to local and foreign investors, for speculation, large-scale oil palm and rice plantations, cashcrops, and other agricultural concessions (Olayinka 2017; Mitman 2021). Land in Africa remains unequally distributed even as equitable service provision remains elusive. Inequalities continue to be on the rise. One of the consequences of the inequities is that political leaders prioritise employment, economic growth, and development which is predicated on the use of the country's bountiful "natural" resources. In Nigeria, this has led to aggressive mining and extraction of oil and mineral from environmentally sensitive areas. Land policies predicated on this line of reasoning continue to imperil the society. High dependence on oil exports has led to skewed exchange rates, the decline of non-resource sectors, political authoritarianism, conflict, and economic inequality. All of these are part of the hidden geographies of the Anthropocene. The colonial subversion of rich indigenous human–environment kinship and substitution with an insidious logic of unrestrained capitalism remade Africa into landscapes of extraction.

Changes in "Yoruba" Human–Environment Kinship[3]

The people known as Yoruba are the second largest language group in Nigeria and one of the most populous ethnic groups in Africa. The Yoruba people consist of several subgroups comprising Oyo, the Ketu, Awori, the Ila, the Owu, the Sabe Ondo, Ife, Egba, Ondo, Ijebu, Ibarapa, Ilaje, Ohori, and Akoko, among others (although with migration, slavery, and globalisation, different forms of the Yoruba groups and language exist worldwide). There are Yoruba groups in the West African countries of Benin Republic, Togo, Gambia, Ivory Coast, and in the Americas in countries such as in Brazil, Cuba, Trinidad and Tobago, and Jamaica) (Akinyemi and Toyin 2017). This chapter focuses on autonomous indigenous groups who shared a similar culture, common ancestral origin, and historical experience and shared a typical origin story as they were confronted with British colonisation in the 19th century.

Before the formalisation and invention of the category of Yoruba as a political identity, internal discord and persistent wars among these indigenous groups stood in the way of a coherent political force (Falola and Oguntomisin 1984). Samuel Johnson's account in *The History of Yoruba* revealed how a declining power of the then mighty Ile-Ife allowed subgroups like Ijebu, Egba, and Ondo to rally around the Oyo kingdom, leading to decentralisation amongst these groups in the 19th century (Johnson 1921). Yet, even as the Oyo Kingdom came to be powerful, it was unable to establish political control over the rest of the region (Falola 1999). When the Oyo Kingdom itself collapsed, wars began amongst other groups (Law 1970). Linguistic and cultural fluidity in the region, therefore, did not have expression in territorial and political terms, nor were political and ethnic frontiers ever stable amongst the groups. The social, economic, and political life of these indigenous groups gave rise to competing empires and complex political formations (Adediran 1994). Perhaps, these indigenous groups were evolving towards a cohesive unit, especially in light of the social, economic, and political turbulence of the 19th century, yet, the political unit – Yoruba – was never in existence. Nevertheless, indigenous human–environment kinship was a common feature among these groups.

Human–environment kinship among precolonial Yoruba indigenous groups are ideologies underpinned by complex systems of divinations that maintained a relationship of reverence and balance with human, the ancestors, and non-human others in the society. The concept of the "environment" or nature as a separate entity from human was not indigenous in many African societies, nor was it Yoruban. Yoruba indigenous groups saw themselves, as enshrined in their origin story and cosmology, as part of and acting upon the world respectful of relationships with the material and non-material beings in the society (Oluwole 1997). According to the late Yoruba priestess and indigenous scholar of philosophy Sophie Oluwole, the Yorubas have a strong belief in the interface between visible and invisible, tangible and intangible as connected in a symbiosis. Land, trees, and forests are not outside of the self, although they have their own agency, they did not assume a defining material category that accords them beneath or above other elements. Rather, human, land, forests, trees, spirits, and animals are amongst the central elements of ecosystem health and balance.

Yoruba gods are known to be affiliated with geographic features including land, mountains, hills, thunder, trees, and forest (Barber 1981). Despite the importance of these categories, they never assumed an overhyped position, but are understood as essential parts of a worldview that promotes balance.

In this indigenous ideology, land was not accorded an elevated status, but like trees, water, and forest bodies it was deified, and selling it was sacrilegious. In 1853, when a missionary enquired about the cost of land from a ruler of Ibadan, the ruler replied in astonishment: "Pay! Who pays for the ground? All the ground belongs to God; you cannot pay for it" (Falola 1984). Land, like others, exists in a symbiotic relationship with other elements and geographic features of the society. Land was a cohesive force that united members of a community. It was administered on behalf of the community by rulers, ancestors, or village leaders.[4] Every family or descent group in the community was entitled to a portion of land and it was through membership of the family that everyone in the community had the right to a piece of land (Akinyemi and Toyin 2017).

Human–environment kinship among the Yoruba can be compared to Botswanan rainmaking which historian and Anthropologist Julie Livingston eloquently described in her book *Self-Devoring Growth*. In precolonial Botswana, chiefs who commanded rain "worked at the interface of an animated ecology that brought frogs, cattle, pangolins . . . clouds, trees, birds, and human past and present among many entities large and small into a dynamic, metonymic relationship" (Livingston 2019: 15). Rainmaking in precolonial Botswana "required that people work in concert, that they respect the plants and animals and clouds and one another" (ibid). Like Livingston described for Botswanan rainmaking, indigenous ideologies are not flawless. Often, leaders use rainmaking to manipulate the society. The 19th-century internecine Yoruba wars show disunity amongst the humans of the regions, and the conflicts are associated with dominion.

A category of the environment in the Yoruba region was the "tree of life"; the oil palm tree. Wild fruits including palm oil used to be public properties and could be harvested by anyone, as they are usually owned communally and could be harvested when ripe, except for those located in people's farms. Palm oil was also a lubricant for drilling tools, and the rounded bead (*ileke*) worn by women on the hip is made from palm-nut shells. Palm oil was also the favourite offering to many of the deities.

Human–environment kinship does not imply that indigenous groups of Southwestern Nigeria do not use the elements of the environment for economic and utilitarian purposes. The Yorubas have it that T'ójú bá ń pón ni, ìgbẹ́ là ń rò fún. This implies that one can consult the wilderness for economic purpose in time of need. For example, the Yoruba have a saying that Kòsí òrìṣà tí kòní ìgbẹ́ which translates as "No Yoruba god without a forest" showing the relationship between igbe or forest and gods. Yet there must be balance and moderation. A Yoruba proverb has it that: Bí a bá ní kí á bẹ́ igi, a óò bẹ́ èèyàn i.e. If one cut a tree, one will cut people. This simply means that one cannot cut trees needlessly and to do so is to destroy humanity. Until palm oil products were aggressively sought after by foreign firms and changed indigenous conceptions of palm oil, it was part of an element of communal ecosocial balance and means of food security.

Palm oil and land will become two "commodities" of great significance for the British authorities, since the transition from the slave trade. Karl Marx *Das Kapital* described "a commodity. . . as an object outside of us (Marx 1988: 45)." Land and palm oil and kernels became commodities and a means of organising colonial life through price regulations, market boards, and labor.[5] Palm oil and palm kernel exports went from few quantities sold to hundreds of thousands in the early 20th century. During the second world war, colonial authorities increased the price of oil and kernels through the West African War Council Supply and Production Committee as conditioned by speculation and demand from the United States and to ease the post-war economic stress of Europe. Colonial authorities deliberated on the impact of price increases on the standard of living of farmers to encourage further production. Palm oil production would become prioritised and elevated under community land management strategies. They became commodities of the British "development" project that feeds capitalism's insatiable appetite. Before the popular Niger-Deltan crises emerged in post-colonial Nigeria, since the discovery and exploration of the region (oil was first discovered in Ondo state towards its border with Ijebu in 1908), conflicts related to land were frequently linked to oil palm plantations. Palm oil exploitation affected the Southwestern region and their rights to forests, livelihood, and culture – the entire balance of indigenous ecosociality.

Land and palm oil commercialisation led to massive deforestation of some of the most biodiverse forests in the West African subregion. Perhaps, palm oil exploitation in Nigeria did not have the conversion of carbon rich peat soils that throws millions of greenhouse gases into the atmosphere, as it did in Southeastern Asia, and as the oil mining in Niger-Delta would come to do. Nevertheless, the shifting idea of land and palm oil commercialisation has its toll on African post-colonial ideas of development. An ideal of progress and modernity is associated with the commercial success of exportation. A capitalist system that gobbles up the present and future of the planet became a post-colonial African vision.

Inventing "Yoruba" in South-Western Nigeria

The idea that power is diffused and embodied in discourse, knowledge, and "regimes of truth" has been sufficiently fleshed out by Michel Foucault (Foucault 1980). In Franz Fanon's analysis of the role of culture and identity in projects of colonial domination, he writes that a key aspect of the colonial project is to ensure that indigenous groups assimilate the hegemonic culture. That, the "native" has to abandon their cultural heritage and assimilate colonial ideas (Fanon 1959). According to the standard narrative, "Yoruba" was a Hausa or Fulani term designating Old Oyo, that was later extended to its vassals and neighbours including Egba, Egbado, Ijesha, Ijebu, Ondo, Ekiti, and Akoko but lacked any overarching identity. The promotion and adoption of "Yoruba" ethnic identity was a precursor to colonialist use of discursive elements to erase indigenous human–environmental kinship. Yoruba as a political entity was different from indigenous southwestern conceptualisation of the self. "Yoruba" was a recent abstraction from these groups, and a political tool that aided the British project and severed human–environment kinship. Yorùbá was reduced to writing by people with very little or no formal training in linguistics or language studies in the early 19th century (Ajayi 1960: 1, 2). In fact, "Yoruba" language only emerged in written form during the 19th century when Samuel Crowther (and CMS missionaries) published a Yoruba grammar book and started translation of the Bible, and was further developed through the efforts of the Yoruba elites – returnee slaves (including Afrobrazilian, Aguda, and Saro) (Falola 1993, 1999).

Interestingly, it was the elite and missionaries that first used the name "Yoruba," to identify these indigenous groups (Falola 1984). While the elites struggled to fit into a rapidly changing Southwestern region, in which they assumed superiority over the indigenous and largely non-Western-trained population, they suffered racial denigration from the British merchants and officials (Falola 1999). The concept of "invented nation" has been used to describe how this Yoruba elite formed a "Yoruba ethnic identity" to build nationalism (Hobsbawm and Ranger 2012). Benedict Anderson has argued that the assertion of group identities through language standardisation and literary production is a key element to the development of a nationalist consciousness. This is evident in the rise of vernacular languages in Europe (Anderson 1983). These Yoruba elites point to a linguistic unity, common ancestral origin, and historical experience as significant factors for the codification of the language. They, therefore, made efforts to standardise the Yoruba language, write Yoruba history and promote the same as a way to forge political identity. During the 19th century, these elites enjoyed considerable interaction with Europeans and mediated between them and the indigenous population (Falola 1999). Having received Western training, they sought to justify local cultures and epistemologies in Western terms. In fact, they shifted the epistemological and ontological premise of indigenous groups and helped realise colonial assumptions about the "African environment."

Two examples of perhaps the most prominent Yoruba elites in Samuel Johnson and Samuel Ajayi Crowther who greatly impacted Yoruba history would suffice. Johnson served as a translator between the Yoruba chiefs and colonial authorities and between Christian missionaries and the Yoruba people in general. He mediated between various Yoruba groups and British leaders during the Yoruba conflicts of the early 19th century (Falola 1999: 31). In his book, Johnson like governor Moloney envisaged a unified Yoruba state, and a Yoruba history that would lead to a conversion of Yoruba to Christianity. Johnson was an Anglican minister and renowned for his pioneer history of the Yoruba people, which remains a standard reference for Yoruba history. In *history of Yoruba*, Johnson has it that "the ancestors of the Yorubas, hailing from Upper Egypt, were either Coptic Christians, or at any rate that they had some knowledge of Christianity" (Johnson 1921: 7). To be sure, while Johnson was writing history, he was also practicing evangelism and offering the services of a pastor. Johnson's effort at relocating Yoruba origin and cosmology was not insignificant. The Yoruba cosmology is the deepest level of hermeneutical vision – where meaning, power, and history converge and where the Yoruba kingdom is ritually made (Apter 1992). While it is true that the Yoruba origin story, remained subject to conflicting analysis, by arguing in favour of a Christian (and European leaning origin) for Yoruba people, Johnson favours the British agenda and ideology (Peel 2016).

Johnson might not be aware that a Yoruba identity that aligns with the British project confounds indigenous human–environment kinship, but his book further indicts him. For example, during his tour around the Ikales (a subgroup of the Yoruba), Johnson described them as a group "inferior intellect" because of their carefree interaction with the land and their lose political system. He described them as "half naked, greasy bodied, dirty and covetous people, occupying a vast portion of land, but living in the thickets without any regular town. No sign of royalty to distinguish them, they are all in their primitive state" (Johnson 1921; Peel 2016: 46). Johnson's denigrating statements all but confirmed the similarity of his views with that of colonialists. He assumed that "vast portions of land" are simply useless and that indigenous communities lacked an understanding of the land and their surroundings. His view, like that of Moloney, is premised on an assumption that the control of the indigenous land system by colonial authorities for colonial exploitation of resources would bring about "civilisation" (read development) to the people.

Similarly, Crowther was famous for translating the bible into Yoruba. Like Johnson, Crowther's project of translations and historical writings demonised ancient gods, contested and dislodged "traditional practices," cultures, and indigenous ontologies. They helped to gradually relocate the centre of the cosmology of indigenous groups. Crowther and other missionaries helped the colonial school systems and churches to teach these new European values. The transformative potential of education cannot be ignored. Colonialists with the aid of missionaries made use of school policy to strengthen the hegemony of the colonial state. These educational and religious policies aligned with the economic and environmental imperative of colonial exploitation. Interestingly, donations for schools were received from palm oil magnate William Lever, the owner of Lever Brothers (Marchal 2008: 3).

As mediators, these elites were doing more than mere translation, they were gradually dislodging forms of indigenous knowledge that do not conform with Western vision and erasing this knowledge through translation. Even later translations – often conducted in close collaboration with or solely by Yoruba people – cannot make up for the concepts that exist in one culture but not in the other. Colonial authorities and later Africans relied on this translation to consolidate their foothold on environmental redefinition and domination. The project of colonialism would never be complete without environmental domination through the erasure of indigenous ecosocial relations. The "Yoruba" that the elite represented and invented became a political tool for nationalism as Africanist scholars have identified. Interestingly, the Yoruba elites' vision of nationalism and modernity, conforms with British authorities, as they viewed and assessed African and indigenous knowledge as "primitive." As scholars have noted, this Western construction of "primitive" Africa will come to influence the rise of "alienated discourse, literacy and the construction of self-identity among Africans themselves" (Bethwell 2009; Mudimbe 1988).

Foremost Nigerian sociologist Oyeronke Oyewunmi has explored the gradual epistemological shift that accompanied translation, discourse, and the imposition of Western notions on gender categories for Yoruba culture as well as its post-colonial afterlives in the production of academic knowledge (Oyewunmi 1997). Within these projects of language, historical appropriations, and translation, lies a violence of erasure – replacing Yoruba cosmology with Western cosmology and human–environment kinship with Western ideals.

Treaties and Territories as Colonial Tools of Environmental Domination

By the end of the 19th century, British colonisers wield formidable power and influence over land use in the Southwestern region. The demand for cocoa and palm oil market amongst others led to the granting of charters and concessions to companies and monopolies like Royal Niger and the Lever Brothers (now Unilever) who helped to extend the colonial reaches of power across the Southwestern region of Nigeria. Market demands and colonial policies shaped investment and agricultural decisions made by farmers, particularly with respect to planting long-term cash crops. It also increased the need for land. Elaborate economies of accumulation and preoccupation with extracting surplus from colonised territories and people produced and consolidated new regimes of land use, in which agricultural systems, in an effort to maximise profits, reconstituted all other facets of life.

Yet these changes would have been impossible without colonial technologies such as treaties and territorialisation. The treaties that the British authorities signed with Yoruba rulers facilitated the formalisation and standardisation of territories and Yoruba political identity as well as the transformation of human–environment relationships. Robert Sack noted that territoriality is the "attempt by an individual or group to affect, influence, or control people, phenomena, and relationships by delimiting and asserting control over a geographic area" (Sack 1986). Colonial authorities divided their colonies into multiple and overlapping political and economic zones, rearrange people and resources within units, and create regulations on how the areas could be used (Vandergeest and Peluso 1995). Territorialising was therefore instrumental in mapping indigenous groups into a geographically circumscribed space of control (Blomley 2019; Stuart 2010). Although Africans contested this remapping and territorialisation, of their landscapes the human–environment kinship ideal was severed.

The colonial treaties generated profound changes that affected indigenous ecosocial relations. From the treaty of Cession (that ceded Lagos to the British authorities) signed by King Dosumu of Lagos in 1861 to those signed by George Tubman Goldie and presented at the Berlin Conference of 1884–1885 to ensure "British occupation" of Nigeria, colonial treaties are more than mere formalities, they give life to colonial assumptions and participate in erasure of indigenous human–environment kinship. A crucial node was the reconstitution of indigenous relationship with land through reterritorialisation and creation of property regime.

For example, when the English tenure system and its individual ownership of land as the system of Crown Grants were introduced and enforced in Southwestern Nigeria, it began a gradual legalisation of colonial assumptions and consolidated the severance of indigenous human–environment kinship. The British ensured that there was a need to get assurance over "land ownership" and to purchase lands and bare titles. It became "imperative that land held under customary tenure must be converted into English tenure, for it to be registered" (Hayford 2003). The idea of land ownership, registration, and permanency of purchase that comes with this law elevated the status of land and created unhealthy competition and rivalry. This is in addition to a new value regime created by largescale commercialisation projects that themselves needed purchasable lands to thrive. This new regime of value would come to alter indigenous ideas forever as indigenous lands became territorialised for the sheer purpose of exploitation of human and natural resources.

It is pertinent here to examine some of the treaties that indigenous kings were coerced into signing and how Yoruba land became territorialised.

> The 1854 Treaty of Epe reads:
> "By the present Document Kosoko Ex-Chief of Epe and formerly King of Lagos do declare that when (as) King of Lagos my *territory* extended to Eastward as far as Palma and Leckie . . . (is) revert to the Lagos government."[6]

In Ketu, another Yoruba province, the British forced the rulers to relinquish their *territories*.

The document reads:

> "We the King designate, Chiefs, Elders and People of the Kingdom of Ketu hereby offer ourselves and our *territory* to be included within the protectorate of her Majesty's Government of Lagos and we hereby declare that our rights and *properties* in the kingdom of Ketu comprise all that *territory* bounded in the North by the country of Barba, on the East by the territory of Alafin of Oyo (Yoruba) from which we are divided by the Awipen- and Ogun River, on the West by Dahomey and the South by Egba, Ilaro, Okeodan and Port Novo"[7]

"Properties" or "territories" in precolonial Southwestern Nigerian region are not conceptualised in the same manner as with Europe as the above treaties would have us believe. Western ideas and perspectives on the spatial organisation are shaped by the concept of property, in which territories are viewed as "commodities" capable of being bought, sold, or exchanged in the marketplace (Soja 1971). It is important to note here that as the British seize these territories, the fate of the non-human-others is left in the hands of the British. In addition, the treaty of cession (that ceded Lagos to the British) has it that "I, Docemo (Dosunmu), do with the consent and advice of my council . . . transferred unto the Queen of Great Britain, her heirs and successors for ever, the port and Island of Lagos." But unlike in England, Dosumu never had the right to cede land (Mann 2007). Nor does land "belong" to human, and accorded such overhyped commercial importance. Casely Hayford sums it up aptly; "by taking away rights of ownership, colonial authorities undermined the social institutions that developed around local management of resources" (Hayford 2003: 3).

Therefore like Moloney's letter, the colonial treaties have the implication of upholding and consolidating colonial assumptions and misreading of African society. Ceding of territories also meant that indigenous authorities lost the right to protect other constituencies of the environment, and gradually, the philosophy helped maintain balance. Delineating and concretising hitherto illegible and sometimes unmarked spaces into British territories meant expunging indigenous ideology about those spaces through discourse. In a bid a consolidate, conquered territories and make "Yorubas one in heart as they are in tongue," colonial authorities severed indigenous-environmental kinship by territorialising shared spaces into one exclusively owned by humans and controlled by the logic of capital. Through these treaties and discourse, colonial authorities created a rigid physical and epistemic boundary for an identity group they helped to create.

Treaties and territorialisation and, the native court system became legal instruments of contest and control within the region. Chiefs became answerable to native courts. Although the British government endorsed the rulings of Yoruban chiefs to the extent that they did not contradict their imperial ideas and judicial practices, they nevertheless overturn the chiefs' and local people's injunctions at their discretion (Mann 2007). Locals often appropriate and upend colonial legal systems for their personal gains, as colonial court records show. Conflicts also emanated within and amongst indigenous groups and British authorities and firms. Many such conflicts involved disputes about differential access and overlap of tenure regimes (Aderinto and Osifodunrin 2012). The conflict and manoeuvring through new colonial policies and value regimes to secure lands for personal use and to respond to a new economic regime of land appropriation and privatisation and intensification of cash crop production reflects an epistemological shift.

Revisiting Land and Land Tenure System Debate

Sara Berry has shown that the seemingly straightforward processes of the commercialisation of agriculture, land, and labor relationships that economists identify as leading the way to greater productivity in agriculture were compromised by the actions of social networks that are rooted in African cultures and history (Berry 1993, 2002). Mahmood Mamdani also noted that the codification of land laws by colonial authorities merely legalised a prevailing trend of land tenure systems in Africa (Mahmood 1996). Others argued that the porosity and duplicity of rights and claims to land and territory doomed a British vision of using land tenure practices to secure economic domination. These scholars view treaties, ordinances, and land tenure systems as having insignificant consequences, albeit for agricultural production and land laws (Ochonu 2013). The argument is also often built upon the notion of Africa as a "land-surplus economy" where land was relatively underpopulated and where there was "hardly a struggle" to experience land issues similar to settler colonies (Hopkins 1986; Bruce 1998).

The previous arguments, however, take Western epistemology and assumption of "African environment" similar to the elites and colonial authorities. The argument measure African growth and development based on the extent to which Africa developed extractive capacities, exploit human and material resources, increased land commodification, and private ownership. Such "progressive" and "developmental" narratives of African history overlook ongoing epistemic violence. These arguments, made largely by scholars of the economic history of Africa, also sought to show that Africans developed or were developing before colonialism, albeit in the Western understanding of development. Buried in discourse and modern ethos, post-colonial imagining of a new African self, and the idea of progress engendered in development projects continues to move away from the African philosophical past and towards a developmental path that Julie Livingston sufficiently termed *Self-Devouring Growth*. This supposedly "forward-looking" idea of growth has "become the organizing logic" "of development," and "its hoped-for future." This development path is an "insatiable growth predicated on consumption" that ultimately overwhelms and is sponsored by a totalising capitalist regime that propels the Anthropocene (Livingston 2019).

Colonial Land Tenure systems prepared the grounds and became a powerful political tool that helped colonialists reconfigure indigenous relationships with land. Land became a separate constituent, in the colonial reconfiguring of the environment. The Yoruba land tenure systems, Colonial treaties, ordinances, and native courts were technologies that gave life to and sutured colonial assumption of a separate nature and society. These transformations were accentuated by the shift in indigenous understanding and relationship to land. In addition to treaties and territories, land tenure systems were a colonial tool that helped colonial erasure of Southwestern indigenous human–environment kinship. By the 1940s, the colonial Land and Native Rights Ordinance held that "the Native Authority of any area in which any land . . . not for public purpose . . . be divided into plots suitable for occupation by natives for business or residentential purpose."[8] Plots of lands were gradually sold by auction, and rents and time of occupancy were set, while logbooks were created to record these processes. Chiefs became rent collectors accountable not to ancestors but to colonial Commissioners of Land, Supreme Court, and colonial authorities through colonial bureaucratic machines; ordinances, treaties, and territories. Compensations and "capital value" of land were determined by yardsticks established by colonial authorities.[9] The land tenure system is the final nail in the coffin for indigenous environmental-kinship. In *The Land Has Changed*, Chima Korieh argued that colonial encounters with the Igbo groups transformed the land and people's identity, as well as the agrarian economy (Korieh 2010). The colonial land tenure system and agrarian change precipitated by colonial extraction and industrialisation incorporated Africa into world capitalist systems as an appendage, and in the process transformed indigenous human–environment kinship and the idea of development through, ordinances, property regimes, and land tenure systems.

This new land and environmental regime allowed British merchants like William Leverhulme to obtain large concessions of oil palm groves in Nigeria (and in the Belgian Congo). In 1935, Leverhulme alone exported 56,000 tons of palm oil and 64,000 tons of palm kernels. While before colonisation, palm produce was for local consumption, alongside Leverbrothers, companies like Leventis, John Holt, and United African Company, and other trade stations would come to dominate its trade. Although the Nigerian Oil Palm Produce Marketing Board (NOPPMB) regulated, large-scale commercialisation and exploitation consolidated new value regimes that favoured the new environmental relationships introduced by British colonial authorities. The colonial governments and later post-colonial governments continue to favor large multi-national companies and commercial capitalists' interests. In the post-colonial Nigeria, the military government continued the policy in 1978 through the introduction of the Land Use Decree (later Act). The Act has it that "all lands comprised in the *territory* of each state in Nigeria are vested in the governor of that state." At the expense of smallholder farmers and pastoralists and communities who were dispossessed of their land, the government sold lands to companies like Shell. Ecological threats like emissions of destructive chemicals, or oil spills from multi-national companies like Shell have prompted a series of resistance struggles by indigenous groups.

Conclusion

In *Out of the Dark Knight*, Archille Mbembe tasked us to see the 20th-century colonial project as part of a planetary reordering of life. Conservation, agriculture, and most importantly, land dispossession and appropriation played a significant role in colonial capitalist projects. Despite African resistance to these processes, exploitation of the environment increased greatly during colonialism as well as the export of surplus that deprived or reduced indigenous societies' forms of the eco-social relationships. In the same vein, resource extraction inaugurated "new cycles of extraction and predation," events that defined the "colonial structuring of economic spaces" (Mbembe 2021: 43). Land, agricultural, extractive, and natural resource mining, industrialisation, and urbanisation, all colonial projects, manifested in radical transformations of the biosphere, with waste, pollution, and epidemic outbreaks in the age of Anthropocene. Across the colonial to post-colonial period, multi-national companies mined oil, timber, and iron-ore creating violent environments (Peluso and Watts 2001).

The epistemic shift follows a long insidious route to environmental violence. Moloney's letter allows us to retrace colonial assumptions and locate colonial tools of environmental domination that shifted the ontological premise of indigenous human–environment kinship. This allows us to understand contemporary environmental crises as historically situated processes and to look beyond spectacular golden spikes for the Anthropocene. Indeed, environmental problems cannot be dissociated "from histories of colonialism, capitalism, and racism that have made some human beings more vulnerable than others to warming temperatures, rising seas, and land dispossession occurring across the globe" (Moore et al. 2019). One cannot account for the origins of the Anthropocene nor trace its genealogy without understanding the role that colonialism played on the African continent during the 19th and 20th centuries. Histories of colonialism allow us to zoom in on the erasure of indigenous ways of life, in the making of globalised commerce, and regimes that reshaped the lives and livelihoods of human and non-human on a planetary scale.

The Yoruba people in their proverbial wisdom posit that "bi oro ba sonu, owe la o fi wa." This roughly translates as, "when we lose direction, the wisdom of our fathers can provide a guide." Wole Soyinka urged intellectuals to regain their indigenous conditions of mind through reeducation aimed at "self-apprehension," and a conscious re-immersion in the endogenous cultural heritage and an act of self-retrieval (Soyinka 1976). Yoruba human–environment kinship viewed in this light offers a lively possibility for rethinking environmental justice in the Anthropocene.

Notes

1 Samuel Johnson 1966 The History of the Yorubas C.SS 509–511.
2 Tamara Gyles-Vernick in *Cutting the Vines* uses *Doli* – the environmental and historical perceptions and knowledge of the Mpiemu (a category of thought in the Central African Republic) to show how communities environmental relationship. Giles-Vernick, Tamara. *Cutting the Vines of the Past: Environmental Histories of the Central African Rain Forest*. (Charlottesville, University Press of Virginia, 2002); See also Paul Richards, Indigenous Agricultural Revolution: Ecology and food production in West Africa (London, Unwin, 1985), p. 41; and 'Ecological change and the politics of African land use', African Studies Review, 26 (1983); Davis Diana K. (2007). *Resurrecting the Granary of Rome: Environmental History and French Colonial Expansion in North Africa*. Athens: Ohio University Press.
3 Yoruba is used to refer to the language spoken by the indigenous groups before and after the adoption of the term 'Yoruba.' It is also used to generally identify the indigenous groups residing in the Southwestern region of Nigeria that speaks this language. I have tried refer to this groups as 'indigenous groups of Southwestern Nigeria' to refer to the period when the identity was not yet solidified in discourse and 'reduced to writing.' I put Yoruba in quotation marks to denote the conflation between the groups and the identity it came to bare.
4 Lagos State Records and Archives Bureau CO583/286/1 Land Policy in Lagos Epetedo Land Ordinance Mevlvyn Tew Commission on Land.
5 National Archives Ibadan CSO26.36183.S.17 Vol II Palm Oil and Kernels.
6 National Archive Ibadan CSO 5/9 (Emphasis Mine) Kosoko who had been granted refuge and given an expanse of land by the Awujale of Ijebu (another Yoruba ruler in the Yorubaland) to rule, relinquished his claims to the ports of Palma and Leckie in Epe.
7 Ibid.
8 Lagos State Records and Archives Bureau CO583/250 The Land + Native Rights Ordinance.
9 Ibid.

References

Adediran, B. 1994. *The Frontier States of Western Yorubaland: State Formation and Political Growth in an Ethnic Frontier Zone*. Ibadan: IFRA-Nigeria.
Aderinto, S. and P. Osifodunrin. 2012. *The Third Wave of Historical Scholarship on Nigeria: Essays in Honor of Ayodeji Olukoju*. Newcastle: Cambridge Scholars.
Ajayi, J. F. A. 1960. How Yoruba was reduced to writing. *Odu* 8:49–58.
Akinyemi, A. and F. Toyin. 2017. *Culture and Customs of the Yoruba Pan African*. Austin: Texas University Press.
Anderson, B. 1983. *Imagined Communities: Reflections on the Origin and Spread of Nationalism*. London: Verso.
Apter, A. 1992. *Black Critics and Kings: The Hermeneutics of Power in Yoruba Society*. Chicago: Chicago University Press.
Barber, K. 1981. How man makes god in West Africa: Yoruba attitudes towards the Orisa. *Africa* 51 (3):724–745.
Berry, S. 1993. *No Condition Is Permanent: The Social Dynamics of Agrarian Change in Sub- Saharan Africa*. Madison: University of Wisconsin Press.
Berry, S. 2002. Debating the land question in Africa. *Comparative Studies in Society and History* 44:638–668.
Bethwell, A O. 2009. Rereading the history and historiography of epistemic domination and resistance in Africa. *African Studies Review* 52:1–22.

Blomley, N. 2019. The territorialization of property in land: Space, power and practice territory. *Politics and Governance* 7:233–249.
Bruce, J. W. 1998. Country profiles of land tenure: Africa, 1996. *Research Papers* 12759, University of Wisconsin–Madison, Land Tenure Center.
Crutzen, P. J. 2002. Geology of mankind. *Nature* 415:23.
Diana, D. K. 2007. *Resurrecting the Granary of Rome: Environmental History and French Colonial Expansion in North Africa*. Athens: Ohio University Press.
Fairhead, J. and M. Leach. 1996. *Misreading the African Landscape: Society and Ecology in a Forest-Savanna Mosaic*. Cambridge: Cambridge University Press.
Falola, T. 1984. *The Political Economy of a Pre-colonial African State: Ibadan, 1830–1900*, 50. Ile Ife, Nigeria: University of Ife Press.
Falola, T. 1993. *Pioneer, Patriot and Patriarch: Samuel Johnson and the Yoruba*. Madison, WI: Program of African Studies.
Falola, T. 1999. *Yoruba Gurus Indigenous Production of Knowledge in Africa*. Africa World Press Trenton, NJ: Africa World Press.
Falola, T. and D. Oguntomisin. 1984. *The Military in Nineteenth Century Yoruba Politics*. Ile-Ife: University of Ife Press.
Fanon, F. 1959. *A Dying Colonialism*. New York: Grove Press.
Foucault, M. 1980. *Power/Knowledge: Selected Interviews and Other Writings, 1972–1977*. New York: Pantheon Books.
Giles-Vernick, T. 2002. *Cutting the Vines of the Past: Environmental Histories of the Central African Rain Forest*. Charlottesville: University Press of Virginia.
Hayford, C. 2003. *The Truth About The West African Land Question Vol 1*. Theommes Press.
Hecht, G. 2018. The African anthropocene. *Aeon. Ecology and Environmental Science* https://aeon.co/essays/if-we-talk-about-hurting-our-planet-who-exactly-is-the-we
Hobsbawm, E. and T. Ranger (Eds.). 2012. *The Invention of Tradition (Canto Classics)*. Cambridge: Cambridge University Press.
Hopkins, A. 1986. The World Bank in Africa: Historical reflections on the African present. *World Development* 14 (2):1479.
Johnson, S. 1921. *The History of Yoruba from the Earliest Times to the Beginning of the British Protectorate*. Cambridge: Cambridge University Press.
Johnson, S. 1966. *The history of the Yorubas*, 509–511. Lagos: Church and School Suppliers Repr.
Korieh, C. 2010. *The Land Has Changed: History, Society and Gender in Colonial Eastern Nigeria*. Calgary: University of Calgary Press.
Law, R. C. C. 1970. The chronology of the Yoruba wars of the early nineteenth century: A reconsideration. *Journal of the Historical Society of Nigeria* 5:211–222.
Livingston, J. 2019. *Self-Devouring Growth: A Planetary Parable as Told from Southern Africa*. Durham: Duke University Press.
Lynn, M. 2002. *Commerce and Economic Change in West Africa: The Palm Oil Trade in the Nineteenth Century*. Cambridge: Cambridge University Press.
Mahmood, M. 1996. *Citizen and Subject: Contemporary Africa and the Legacies of Late Colonialism*. Princeton: Princeton University Press.
Mann, K. 2007. *Slavery and the Birth of an African City: Lagos, 1760–1900*. Bloomington: Indiana University Press.
Marchal, J. 2008. *Lord Leverhulme's Ghosts: Colonial Exploitation in the Congo*. Translated by Martin Thom, with an introduction by Adam Hochschild. New York and London: Verso.
Marx, K. 1988. *The Communist Manifesto*. New York: W.W. Norton and Company.

Mbembe, A. 2021. *Out of the Dark Night*. New York: Columbia University Press.
Mitman, G. 2021. *Empire of Rubber: Firestone's Scramble for Land and Power in Liberia*. New York: The New Press.
Moore, S. et al. 2019. Interrogating the Plantationcene. *Edge Effects*. https://edgeeffects.net/plantation-legacies-plantationocene/
Mudimbe, V. 1988. *The Invention of Africa*. Bloomington: Indiana University Press.
Nixon, R. 2013. *Slow Violence and the Environmentalism of the Poor*. Cambridge, MA: Harvard University Press.
Ochonu, M. 2013. African colonial economies: Land, labor, and livelihoods. *History Compass* 11 (2):91–103.
Olayinka, A. 2017. The intersectionalities of land grabs in Nigeria: Engaging the new scramble for African lands. *Africa Insight* 46 (4).
Oluwole, S. B. 1997. *Philosophy and Tradition*. Lagos: African Research Konsultancy.
Oluwole, S. B. 2014. *Socrates and Ọ̀rúnmìlà: Two Patron Saints of Classical Philosophy*. Lagos: Ark Publishers.
Oyewunmi, O. 1997. *Invention of Women: Making an African Sense of Western Gender Discourses*. Minneapolis: University of Minnesota Press.
Peel, J. D. Y. 2007. Olaju: A Yoruba concept of development. *The Journal of Development Studies* 14:139–165.
Peel, J. D. Y. 2016. *Christianity Islam and Orisa-Religion: Three Traditions in Comparison and Interaction*. Oakland, CA: University of California Press.
Peluso, N. and M. Watts. 2001. *Violent Environments*. Ithaca: Cornell University Press.
Richard, P. 1983. Ecological change and the politics of African land use. *African Studies Review* 26:1–72.
Richards, P. 1985. *Indigenous Agricultural Revolution: Ecology and Food Production in West Africa*. London: Unwin.
Sack, R. 1986. *Human Territoriality: Its Theory and History*. Cambridge and New York: Cambridge University Press.
Saïd, E. 1993. *Culture and Imperialism*. New York: Pantheon Books.
Soja, E. W. 1971. The political organization of space. *Association of American Geographers Commission on College Geography Resource Paper* 8. Washington, DC.
Soyinka, W. 1976. *Myth, Literature and the African World*. Cambridge: Cambridge University Press.
Stuart, E. 2010. Land, terrain, territory. *Progress in Human Geography* 6:799–817.
Vandergeest, P. and N. L. Peluso. 1995. Territorialization and state power in Thailand. *Theory and Society* 24:385–426.
Wande, A. 1976. *Ifa: An Exposition of Ifa Literary Corpus*. Ibadan: Oxford University Press.
Watts, V. 2013. Indigenous place-thought and agency amongst humans and non-humans (first woman and sky woman go on a European world tour!. *Decolonization: Indigeneity, Education and Society* 2:20–34.
Wayne, Y. and E. Tuck. 2012. Decolonization is not a metaphor. *Decolonization: Indigeneity, Education and Society* 1:1–40.
Whyte, K. 2018. Settler colonialism, ecology, and environmental injustice. *Environment and Society: Advances in Research* 125–144.
Yusoff, K. 2018. *A Billion Black Anthropocenes or None*. Minneapolis: University of Minnesota Press.

8 A Revaluation of Traditional Ecological Thoughts, Knowledge, and Practices of the Aari of Southern Ethiopia

Endalkachew Hailu Guluma and Zewde Jagre Dantamo

1. Introduction

Colonialism/neo-colonialism and globalisation have made modernisation in Africa to be considered synonymous with Europeanisation. The western distinction between tradition and modernity is "a method of indoctrination aimed at dismissing non-occidental pasts as outdated" (Kebede 2001 quoted in Salvador 2007: 561). The failure of many African countries is associated with their inability to take indigenous values as anchoring point in their reconstruction and development programmes by a plethora of scholars (Verhelst 1990; Gyekye 1997; Serequeberhan 1997; Teodros 2001; Kaphagawani and Jeanette 2003; Kebede 2004; Kenenisa 2010; Akpomuvie 2010). Wole Soyinka has called for revaluation or "the selecting and lifting up of the deeper (viable) traditional values of a community and recasting them to meet today's realities" (Soyinka 1998: 89). Similarly, Gyekye (1997) calls for the appreciation of traditional values as a way of reviving our vital values. Similarly, Kenenisa (2010) argues that a revaluation and revival of traditional indigenous African values and knowledge, coupled with the appropriation of modernity by contextualising it and rejecting its universalistic conceptions, is required in Africa (Kenenisa 2010). Similarly, wa Thiong'o (2013) calls for a return to African elders who Oruka (1991) says practice "sage philosophy" to develop a genuinely African education, knowledge, and philosophy by marrying the traditionally oral knowledge-making in folklore (tongue) with the modern (pen) (wa Thiong'o 2013: 160). In general, a critical revaluation and selective appropriation of modernity and tradition in all realms of life/knowledge seem to be what African philosophers and scholars are generally calling for.

One such realm that needs revaluation and appropriation is ecological knowledge. African ecology and conceptions of ecology have suffered colonial violence and continue to suffer under global capitalism and western cultural and epistemic hegemony. According to Edward Said, "imperialism is an act of geographical violence through which virtually every space in

DOI: 10.4324/9781003287933-8

the world is explored, charted, and finally brought under control" (Said 1978: 77). The process of decolonisation must, therefore, involve reclaiming both the physical landscape and the native community's cosmology and ontology that constitutes and gets constituted by it. In this regard Said states "the history of colonial servitude is inaugurated by the loss of locality to the outsider; its geographical identity must thereafter be searched for and somehow restored" (Said 1978: 77). It is not only for the purpose of decolonisation that this needs to be done according to Ikuenobe (2014), but also because "the activities that have raised environmental concerns in Africa did not exist prior to colonialism because Africans had conservationist values, practices, and ways of life" (Ikuenobe 2014: 2). According to Ikuenobe "nature is seen in traditional African thought as holistic and as an interconnected continuum of humans and all natural objects which exist in harmony" (Ikuenobe 2014: 2). Traditional African people's activities and lifestyles showed "their efforts to exist in harmony with nature" and resulted in the conservation of nature (Ikuenobe 2014). As a solution, Ikuenobe (2014) suggests that Africans should unlearn colonialism's biases that generally privileged western tradition over the Africans. At a time when "contemporary environmentalist views and movements represent traditional African beliefs, ways of life, and moral views that have been rejected by Europeans" in the past, such an unlearning accompanied by a revaluation of traditional African views about the environment plays a vital role (Ikuenobe 2014: 20).

Concerning the environment, what should be unlearned are the colonial structures inherited from the era of colonialism and sustained presently in the name of modernity and globalisation and the resultant "ecological imperialism" exacerbating environmental problems in Africa (Tiffin 2007 cited in Huggan and Tiffin 2010: 3). Tiffin (2007) calls the overall effect of European colonial enterprise on the environment and the non-European subjects and spaces "ecological imperialism" following Alfred Crosby (1986) (Ibid.). Ecological imperialism continues to dictate ecological knowledge/sciences/discourse to this day with tacit speciesm that defines human in the image of the European male and the non-male European and nature/animal as not-human empty space (Plumwood 2003: 53 cited in Huggan and Tiffin 2010: 5). Plumwood argues that it is such Eurocentric anthropocentrism that justified European colonialism of indigenous cultures as primitive (Ibid.). In this light, it is not only the European conquest and global domination that exacerbated ecological imperialism but the ideologies of imperialism and racism inherently carry with them ecological imperialism (Huggan and Tiffin 2010: 6). Post-colonial ecologies still continue to be affected by such ecological imperialism due to global capitalism and the west's continued role as the centre of ecological discourse and knowledge. Building on these insights, this study re-evaluates traditional practices and proposes a selective appropriation and synergy of the traditional and modern using community-led conservation models.

2. The Case of Aari People

Ideologies of imperialism, which inherently include aspects of ecological imperialism, are not limited to the global or continental level. At national levels, the culture and knowledge system that historically becomes hegemonic often imposes its own ecological imperialism, which is allied to global imperialism. The mainstream Ethiopian and western views of the environment have together disenfranchised traditional ecological knowledge in multi-cultural Ethiopia including those of the Aari people.

The Aari people live in South Omo Zone's North Aari and South Aari districts. Traditionally, the Aari people used to live under nine self-governing territorial groups with each having its king/*babi* (Yntiso 2010: 186). Yntiso further explains that "*babi* was advised or assisted by councils of elders (*galta*), chiefs of sub-divisions (*ganna*), ritual specialists (*godmi*), and information agents (*tsoyki*)" (Yntiso 2010: 186). The nine groups are "Baka, Kuramer (Kure), Shangma/Layda, Sidamer (Sida), Beyamer (Beya), Wobamer (Woba), Gayil, Bargid, and Seyki or Argen" (Yntiso 2010: 186). King Menelik II incorporated them into the modern Ethiopian state at the end of the 1880s (Naty 1994: 500). Following their incorporation, central Ethiopia's administrative structures and Christianity gradually infiltrated

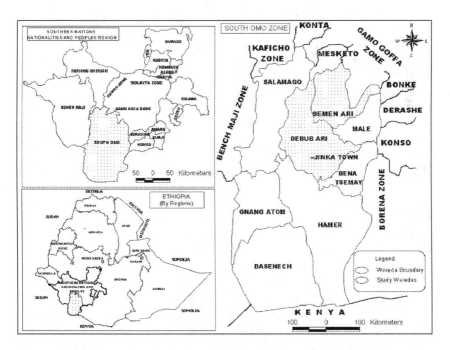

Figure 8.1 Map of Aariland (areas indicated as Semen Ari and Debub Ari).
Source Yntiso (2010: 184)

the self-governing territorial groups. The entry of the Ethiopian Orthodox Church began with soldier settlers building churches and small towns. This was followed by the passing of a decree on forceful conversion by the Haileselassie regime in the 1940s, and the arrival of protestant missionaries from Sudan in the 1950s (Naty 2005; Yntiso 2010). Incorporation also led to the expansion of gäbarə sərə 'ätə (feudal system) in which former dignitaries of the Aari people and some soldier settlers from the centre became feudal lords (*balabatə*/landlords or estate owners), and the rest of Aari people became *gäbar* (tenants) paying taxes to them (Naty 1994: 504–505). An unpublished study report of South Omo Zone's Culture and Tourism Department (SOZTD) entitled *"yä 'äri bəheräsäbə 'ät'äqalayə gäs'ətə" (General Overview of Aari People)* claims that three-quarters of the land formerly under *Babis* and communal ownership was placed under the administration of the new *balabatəs*/landlords, who became the emperor's agents (SOZTD n.d.). It further documents that in some places some dignitaries and newly appointed *balabatəs* ran the King's administration while *Babis* ran parallel traditional administration. The agricultural landscape thus gradually began to change and the new elites in the state structure and Orthodox and Protestant Christian missions became new centres of power and knowledge.

Following the 1974 revolution, the Marxist military regime of the Dergue further strengthened the central state administrative structures placing all rural and urban Aaris under centralised administrative units overlooked by semi-literate and literate Dergue appointed elites. The few former *Babis* that used to be feudal landlords were bereft of what was left of their powers as land was distributed to peasants. All religions including traditional religion were invalidated as irrelevant ideological indoctrinations of feudalism (Karbo 2013: 46). The administration was highly centralised and the environment and people continued experiencing ecological imperialism.

Although the EPRDF promised to empower ethnic groups and decentralise administration when it came to power in 1991, it turned out to be a centralised government with centrally planned policies (Weis 2015). Although aspects of the Aari culture and identity began to be celebrated, the modern Aari political and Christian elites together with literate Aaris under their influence (and under both national and global centres of power and knowledge) have continued to disenfranchise Aari traditional intelligentsia's indigenous ecological knowledge by knowingly/unknowingly perpetuating ecological imperialism.

In spite of this, various studies such as those by Tolessa et al. (2013), Kidane et al. (2014), and Kidane et al. (2015) show that the Aari people have preserved plant biodiversity in their areas for centuries. But they are currently threatened by agricultural expansion, overgrazing, and drought (Tolessa et al. 2013; Kidane et al. 2014). A new revitalised move towards biodiversity conservation of the rich plant species in Aari land is necessary (Kidane et al. 2014; Kidane et al. 2015). Though studies have acknowledged the preservation of plant biodiversity, medicinal plants, and their eco-friendly "enset" farms (Yntiso 1996; Shigeta 1990; Tolessa et al. 2013; Kidane et al. 2014, 2015), they insufficiently examine the Aari indigenous knowledge and folklore that has led to biodiversity conservation. Most Aari still follow aspects of their traditional religion, but it should be underlined that this includes those who practice rituals of the traditional religion despite being converts to Orthodox Christianity and Protestantism (Encyclopaedia Aethiopica 2003, vol. 1, 3). Aari people believe that there are supernatural gods Sabi (a male god), Bäri (a female goddess), and ancestral spirits ('aka/ 'aki/bunəka). Everyone is believed to have his own Sabi and Bäri (Yntiso 2015: 44). The offerings to these deities are given "in the holy groves or in the granges of chiefs and clan elders" (Encyclopaedia Aethiopica 2003, vol. 1, 3). The Godəmis possess a hereditary priesthood in Aari traditional religion and have the power to depose the Babis in case of inability to fulfil their obligations (Encyclopaedia Aethiopica 2003, vol. 1, 2). But this traditional religion's role in environmental conservation is yet to be studied. Indigenous knowledge about the environment and human–environment relations in Aari's 'ata keza (myths, legends), traditional religious beliefs, and rituals also remain understudied (Kidane 2015: 51–52). With a new asphalt highway project presently cutting through Aariland, attracting more trade, investment, and urbanisation, it is important that we begin thinking of ways of salvaging what is left of their traditional ecological knowledge.

The data collected through observation and unstructured snowball interviews of Aari elders and *Godəmi*s are analysed and discussed below to find out Aari traditional ecological thoughts, knowledge, and practices in respect of how eco-friendly and pro-conservation they are, the extent to which they are threatened, and how they could be recuperated to augment conservation in the modern age.

3. Analysis and Discussion of Results

The results of the analysis of data collected through unstructured snowball interviews with Aari elders clearly indicate that the environment in Aari traditional thought is a holistic place encompassing humans, spirits, and deities. They live together in it with plants and animals, communicate, and visit each other. Humans join their dead ancestors' spirits in this same environment on earth when they die, and gods and ancestral spirits constantly visit, watch over, and live with humans inside the environment. Only *Bäri* and *Sabi* are believed to reside in the sky, but they are believed to constantly visit (reside with) their creatures. Living with deities and ancestral spirits is guaranteed as long as each household puts aside an *'aka* grove behind the house as a place of worship. Each clan must set aside an *'aka* forest, and keep and respect the sanctified groves, forests, springs, trees, and hills assigned to the spirits and deities. When one lives with spirits and deities like this, one lives under their grace and blessing with plentiful harvest and natural balance (sufficient rainfall, rivers, and fertility). The environment in Aari traditional thought is, thus, not the sole property of man but is of the spirits and deities too. Not all trees, forests, or springs are consumable because some are given soul and sanctity, and are associated with the spirits and deities. But there are homestead farmlands (*tika haami*) where root crops, fruits, and vegetables are grown, crop fields further away from home (*wony haami*) for cereals, pulses, coffee, and cardamom, and common lands, grazing lands, rivers, and springs which can be used. Clearly, the Aari tradition conserves the environment through a wise and balanced usage. These findings concur with those by Shigeta (1990: 94), Yntiso (1996: 108), and Noguchi (2013: 139). The Aari understand the interrelatedness of the natural elements. For them, as long as the sanctified forests, groves, big trees, and springs are kept, rainfall will come in time, yields will not fall, and rivers will not dry. The environment does, therefore, not only comprise animals and plants, but is completed by the presence of gods, spirits, and humans, in interaction with one another. The earth or natural environment is under the custodianship of gods who visit and live in it, and through which humans are created and return upon death. It, therefore, commands respect. A more detailed analysis and discussion of the Aari traditional religious beliefs, rituals, and myths that embody them follows.

3.1. Sanctified Groves, Forests, Springs, and Places of Origin

According to Aari traditional religion, man's role is to keep the environment in which he and the spirits of his ancestors live, and in which the gods *Sabi* and *Bäri* often visit. The places they reside in when they visit are called *'aka/'aki/bunəka* (ancestral spirit's) groves. As highlighted earlier, every family must maintain an *'aka* grove behind the house of a family elder. There should also be a clan *'aka* sanctified forest. When a *Godəmi* elder called *Sabibabə* informs the elder of a family or *Babitoyədə* of their arrival, he will offer sacrifices to them in the family grove. Seasonal sacrifices are given by the clan's *Godəmis* in the holy forest of the clan. In return, they will be blessed until they finally join their ancestors' spirits and start living with them. The spirits of the dignitaries of Aari people like *Babis* and *Godəmis* are believed to join their ancestors in the sanctified groves and forests. But the spirits of *käyəsi* or commoners are believed to go to a place called Karo near the Omo river in the valley. The spirit world in which ancestors live and the physical world in which those alive live are thus found in the same environment. The environment thus becomes both their place of origin from which *Sabi* and *Bäri* created them and their final abode as spirits. Followers of Aari folk religion thus cannot say "I'll leave earth and go to heaven when I die. Why do I care?!" like followers of modern religions. This belief clearly contributes to respect and care for nature. With such respect, the Aari believe in keeping the earth as a complex and respectable abode of humans, spirits, deities, plants, and animals, making them not to consider the environment a consumable and destructible property of humans only, unlike modern anthropocentricism.

In this study, we were able to interview two *Godəmi* elders (in Phelpha county around Metser and in Sanmamer county around Senegal) who inherited traditional religious powers and roles from their fathers and who protect groves/forests their ancestors hailed from. Both had small-protected groves behind their houses, which cannot be entered by any human being and livestock. Only the *Godəmis* themselves can enter them to perform traditional religious rituals. We have seen these groves and witnessed that they have big indigenous trees and dense undergrowth untouched by both humans and tamed animals. Aari farmers living around these groves/forests, even the Christian ones, swore they won't dare touch or enter them because they respect and fear the power of the deities, spirits, and the *Godəmis*. There is an *aka* spring in the forest behind one of the *Godəmis'* house in Sanamamer county that is sanctified and used only for ritual purposes. The *Godəmi* and those who share his belief consider drinking water from this spring as drinking their ancestors' blood. As a result, the river and water systems are protected. The *Godəmis* and their followers believe that if the groves/forests and springs are destroyed, this could reduce rainfall, fertility, and plentiful harvest.

The *Godəmi* in Sanmamer county around Senegal has a bigger forest he protects in addition to the grove behind his house. He is highly venerated and has the title of *Gudurubabi* (literally lord of the hyenas). He has a special power over hyenas, which, it is believed, he can order. Whenever anyone or anyone's livestock touches the forest he protects, hyenas will visit and attack that person and/or his livestock at night. It is only he who can perform a ritual that can keep away the hyenas when the trespasser confesses to him and repents. Due to this, people will not enter the forest he protects. The *Gudurubabi* is also highly venerated and respected because he is considered a person upon whom *Sabi*, *Bäri*, and ancestors' grace resides. Saddening him or *Sabi*, *Bäri*, and ancestors by touching the forests is believed to result in curses and calamities.

Though the government's declaration of the bigger forest as a protected forest may have assisted him in his job, no incentive, recognition, or payment of any sort has ever been given to him. The *Gudurubabi* protects the forest out of commitment to his worship and veneration of *Sabi*, *Bäri*, and ancestral spirits. He inherited the responsibility and power of protecting the forest from his father and will pass it on to his first son, and the protection will continue. In our view, the approach used by traditional keepers like him may yield better conservation outcomes compared to the government models that tend to be inconsistent and inefficient. We think that traditional keepers of forests should, therefore, be empowered and supported by introducing community-led conservation as this not only will provide a more permanent and consistent environmental conservation but also helps preserve the indigenous eco-friendly ecological thoughts, knowledges, and practices. Well-designed community-led/based conservation projects are proven to succeed in communities where there are supportive cultural beliefs and institutions according to Brooks et al. (2012). Numerous other scholars have called for community-led/based conservation in similar contexts (Lepp and Holland 2006; McAlpin 2007; Horwich et al. 2011; Chung-Tiam-Fook 2011; Chilly 2017; Turreira-García et al. 2018).

Aari traditional religious beliefs, myths, and rituals commonly consider the environment as the origin or source of life and as holy. This is because most Aari clans believe they were created from a certain natural element in their ancestral place. For example, the clans called *Woč'a* and *'amänə* believe they came from the earth while the clan of *Binəkə* is believed to have come from water. The clan of *Gadetə* is believed to have come from the sun. A clan around Wobamer believes that it and all humans were created from a giant bamboo found in a gorge in Wobamer. There is a clan believed to have come from a giant Lagenaria gourd. A clan called *'arəma* came from a mountain. And the list continues. Some highly regarded *Godəmi* clans came from Arki in Sido and Sokelet in Wobamer which are highly venerated holy forests where the rituals of coronating kings/*Babis* used to take place. The Ashti forest of Bako and the Wogezen forest of Biyo are also considered the places where *Godəmi* and *Babi* clans hail from. We visited the Ashti forest and found it to be an untouched dense natural forest with lots of giant indigenous trees, dense undergrowth, lots of birds, few wild animals, and a river that cuts it into two. It has preserved the plant biodiversity of the area as we are told that rare medicinal plants are to be found there. These natural environments (forests, mountains, springs, hills, rocks, trees, plants, etc clan founders are believed to have come from) are sanctified and conserved for centuries. In the past nobody, except *Godəmis*, would dare to set foot in them. But nowadays, albeit being gazetted as protected, people now regularly trespass into forests, especially where hereditary keepers of the forests are no longer available (mostly due to conversion into Christianity). We further observed that there are low-scale logging activities taking place in these forests. The shift of the centre of power and knowledge from traditional Aari intelligentsia to political administrators and Christian elites meant elders won't be heeded. In addition, officers assigned as administrators and forest development and protection officials are inefficient. As one elder put it "it is not illegal logging when they cut trees, but it is illegal logging when we cut it." This clearly shows how corrupt, alienating, and ineffective the officials are compared to the traditional keepers. This is why we argue that the powers and responsibility of conservation should be ceded to the community and traditional keepers (like *Godəmis* and *Babis*), who should then be involved in a community-led conservation for a sustainable and effective conservation.

3.2. Huge Trees

Traditionally, the Aari people of the past rarely cut down huge trees from the bottom of their trunks even in non-sanctified areas. They cut the branches of trees. Big trees are believed to be filled with the grace of the deities and the ancestral spirits and respected. They are considered as having souls/lives. Therefore, it is not uncommon in the Aari language to refer to the cutting of trees as "killing of trees." Elders explained that the person who cuts a huge old tree down is believed to die soon by divine punishment in the past. So, cutting down old huge trees is not done except after consulting *Godəmis* and *Babitoyədəs* in exceptional circumstances. But nowadays, people can plant replacement seedlings, show them to the *qäbäle*/county officials, and be allowed to cut the trees down as long as they are not in protected/sanctified forests or groves. This is clearly not as effective because it will only result in a commercial planting and cutting down of young trees for material gains. It is clearly a modern Christianity and European science based on ecological imperialism because it changes the meaning or ontology of trees among the Aari people. Trees will gradually lose their rich traditional meaning and become objects anyone can cut down and sell for cash resulting in deforestation instead of conservation. We think that involving Aari elders, *Godəmis*, and descendants of *Babis* in a community-led conservation will help preserve both the trees and their rich traditional ontology.

3.3. S'oyəsi Forests

The reason why people will not touch some forests is not only due to respect for *Sabi*, *Bäri*, and ancestral spirits but also due to fear of *s'oyəsi*. In Aari traditional religion, *Sabi* and *Bäri* are believed to send spirits that attack people called *s'oyəsi* for various reasons. According to Aari elders, these spirits are not the same as what modern religion calls Satan or evil spirits. The spirits are believed to reside in some forests. So, if anybody, out of his free will, enters such forests, he will be attacked by them. Before a person goes into such forests, he should consult *Godəmis* and the elders of the clan to seek permission. The *Godəmis* will perform a ritual that makes the person immune to the *s'oyəsi* spirit. Nobody collects firewood from the forest and tamed animals will not enter it. As a result, *s'oyəsi* forests were preserved for centuries in Aari land. *S'oyəsi* forests are also seen as keeping harm away from the residential and farm areas where people live. For instance, when a person suddenly gets sick and won't get better, the Aaris believe that he is attacked by *s'oyəsi* because he has done something wrong. A *Godəmi* will come and perform a *s'oyəsi* removal ritual. The *Godəmi* will go into a particular place in the forest where he will leave what he has collected from the compound of the sick person during the ritual. Immediately the *s'oyəsi* spirits will leave the sick person for the forest in the form of a shooting star shooting from the grove behind the home of the sick to the nearby *s'oyəsi* forest. The sick person will be miraculously healed. These rituals seize the community with fear, yet they still need the *s'oyəsi* forest to stay. Without it, they will be exposed to the harmful *s'oyəsi* it has kept at bay. The government can facilitate community-led conservation of such *s'oyəsi* forests, which at some places have become targets of destruction by Christian fanatics.

3.4. Respect/Power, Holiness, and Nature

Aari myths associate respect, power, and holiness with nature. There are myths that are told about the origins of high-ranking and respected *Babis*. For example, there is a myth that

> Once as people were gathered for their own purpose in Bako's centre Ashti, a woman called *Gäli* descended from the sky and landed on earth. This woman held a dried bamboo log in her hand. Without talking to anyone she called *Sarəsarə* who was found by *Godəmis* in Ashti forest in the *womä'a* (sycamore) tree. *Sarəsarə* has hands and legs of a human, face like a baboon, never speaks, and eats only *womä'a* (sycamore) tree's fruit. When *Sarəsarə* came to *Gäli*, she gave him a dry bamboo log that many several people sought. *Sarəsarə* threw the dry bamboo log onto the ground and, immediately it grew leaves and branches. This man became the father of *Babi* Masa and, the *Babi* of Baka. The Masa and Baka royal residences are in Ashti forest.

To this day, the species of this bamboo is not used for any purpose among the Aari of Baka and is referred to as *Bäri /Sabi-'oyəsə* or the bamboo of *Sabi/Bäri*. It is only used for hanging Lagenaria gourds (*leqəsi-'oyəsə*) when a *Babi* dies.

In the previous myth, the nobility of *Babis* of Baka is associated with the forest, tree, and fruit-eating unique human living alone in the Ashti forest – *Sarəsarə*. Clearly, the forest, the tree, and its fruit are associated with purity because the Ashti forest is a sacred forest where the ancestors of *Godəmis* hail from. Thus, by associating their origin with it, the *Babis* of Ashti are associated with purity and venerated as saints. *Sarəsarə* was selected over people living in villages because he lived in harmony with nature. He was, therefore, considered to be free from sin. To this day, some *Godəmis* avoid small towns and villages to remain holy and preserve their spiritual powers. A powerful *Godəmi* around Tembel we were unable to contact is a good example of this. The myth also says, *Sarsar* threw the dry bamboo log and it immediately grew and sprouted into leaves suggesting that he was blessed. Among the Aari people, a person is regarded as blessed when what they plant grows. Power and sacredness of the *Babis* of Ashti come from the closeness with nature they inherit from *Sarsar*. The *Babis*, giant *womä'a* / sycamore tree, and the Ashti forest itself are respected by the Aari people. The *Sabi/Bäri*'s bamboo species are, in turn, conserved.

The beliefs in this myth and the historical fact that Ashti is the royal residence of the *Babis* of Baka, who led their people for years and defiantly fought against Menelik II's forces (Naty 1994), makes Ashti and its forest a place of great significance for the Aari people. Thus, community-led conservation and ecotourism could preserve the forest and provide alternative means of income that can flourish in this area. If a traditional royal house is built at the place where the royal house and compound of *Babi* Masa or *Babi* Mugjobaiso used to be together with a museum and library displaying Aari

A Revaluation of Traditional Ecological Thoughts 133

cultural artifacts, heritages, and history, it will attract tourists generating income for the community of the area and will also help pass on their historical heritage and traditional ontology of environment to the next generation.

The criteria used for choosing a *Babi* successor depends on living in harmony with nature. For example, a successor will only be chosen if bees have built their hives in the house in which he was born and raised or if a bird called *kəqesə*, easily frightened by people, lays her eggs on the grass of his roof. These seem to be indicators that he can live in harmony with nature and has a mythical connection with it. The bees won't bite him or anyone he lives with. The *kəqesə* bird will not be scared of him. He is also expected not to have eaten chicken and eggs. In addition, he is expected from clans under the moiety of *'äšänəda*, which have come from and have powers over different natural features, e.g., *woč'a* has power over Earth and *gadetə* over the Sun.

Babis are tasked with solving natural disasters related to drought, over-flooding, persistent rain/snow falls during the wrong season, disease epidemic, locust and bird invasions, etc. Babi performs rituals during which he consults the Godəmis and elders to solve the problem. If they fail to resolve the problem, Babi will consult with renowned Godəmis and elders. He will pray and slaughter a sheep, read its intestine, and give orders to the community on the issue. So, having power is defined in terms of living in harmony with and having certain mythical/spiritual power of predicting nature and guaranteeing natural order in consultation with ancestral spirits and deities. Babis and Godəmis, the most powerful members of the Aari community, are appointed and respected based on their connection to nature. All of these illustrate how closeness to and living in harmony with nature are highly valued in the Aari tradition unlike the modern religion and western science-based anthropocentric ecological imperialism, which are imposed upon them.

3.5. Green Leaves and Shades

In Aari rituals, myths, and traditional religion green leaves are considered clean and holy. When a *Godəmi* or *Babitoyədə* is planning to perform a ritual for *Sabi*, *Bäri*, or ancestral spirit, he is expected to sleep on fresh green leaves of *enset* (*Ensete vetricosum*), wake up the next morning, slaughter a sheep inside the grove, read its intestine, and serve its meat on an *enset* leaf. The leaf of the *enset* upon which the sacrificed meat gets served symbolises purity and cleanliness, which, according to Aari belief, is only found in pristine nature. The green *Sakətə* grass, used in newly born babies, newly wedded couples, and various kinds of cleansing rituals, is also highly regarded. It symbolises natural purity and fertility in the hands of *Godəmis*. The following mythical belief even extends the symbolism of the leaves of this grass to holiness and antidote of evil:

> When an Aari person embarks on a long journey across many fields and forests, it is believed that he will be protected by shades of trees from evil lurking in the sky. When a flying eagle casts its shadow on a person,

it is believed that the person will become sick. So, when walking on a path without trees, a person is encouraged to place a *sakəta* grass leaf over his head. This single grass leaf is believed to protect him from the evil lurking in the sky.

In the previous myth, the shade of trees and a grass leaf are shown as having the power to repel evil. Living in densely forested areas under the shadows of trees is, thus, considered protective and desirable compared to living in an open deforested field. To this day in Aariland, many compounds are surrounded by various fruit-bearing or non-fruit-bearing trees. So, the presence of green leaves and trees in Aari communities symbolises safety and protection. This belief has helped preserve lots of plant biodiversity in the area. Current conservation efforts in the area can build on such beliefs in the people's communal unconscious.

3.6. Homegrown Calendar, Seasons, and Agro-ecological Zones

The Aari people have indigenous ways of marking time, months, seasons, and agricultural zones. They follow a lunar calendar, and a year is believed to have four seasons. The first season, *Häšänə*, is a sunny and harvest period and is from mid-November to mid-March. The second, *Wäli*, is from mid-March to May. *Bärəgi* is the rainfall season and is from June to August, and *šəkuli* is from mid-September to mid-November. *Wäli* and *šəkuli* have a balance of warmth and moisture. The Aari also uses the new moon and the direction of its narrow tip to predict seasons of drought and seasons of plentiful harvest. This division of seasons is based on natural phenomena in their area and helps them predict sowing/planting and harvesting times. Knowledge of the appropriate time for sowing seeds is a valuable asset key to the survival and success of the Aari people as the following proverb indicates.

> *Phute gaysaq'ab ken deysa kul'ensde, phultenk shedaq'ab ken gaba phurintde.*
> The wife of the farmer that sowed seeds at the right time grinds grains, but the wife of the one that didn't runs to the market.

Every month starts when a new moon first appears and ends when it disappears. The number of days of the months are from 28 to 30 and a year has 12 months (*Soməza, Lonəga, Tora, Makanə, 'äyədi, Dola, La, Täbəza, Zela, Sagəla, Tama, Bäkəsa*). Aari people do not count years and there is no tradition of the first month or new year (*General Overview of Aari People* n.d.). But because during November/ *Soməza* there is a tradition of celebrations and feasts from the new harvest, *Dəšəta* – thanksgivings that take place in the houses of *Godəmis* and *Babis* first, and later in the month of December/ *Lonəga* among the commoners – there is a tradition nowadays of considering this season as the time of the entry of the new year and celebrating it (SOZTD n.d.). The Aari also traditionally divide their land into *dawla* (hot lowland), midland (no local name), and *dizi* (cold highland), and have over the years acquired indigenous knowledge on the understanding of these agricultural zones as well as appropriate crops and livestock for each zone. This finding supports the findings by Shigeta (1990: 94), Yntiso (1996: 108), and Noguchi (2013, 138–139). Concerned stakeholders shall promote and conserve this indigenous calendar, knowledge of seasons, and agricultural zones as they are significant to their eco-friendly environmental ontology.

3.7. Threats to Aari Traditional Ecological Thoughts, Knowledge, and Practices

Based on our interviews with elders and field observations, we identified four major challenges to the eco-friendly traditional ecological thoughts, knowledge, and practices in Aariland. The first is centralised administration by Christian elites. Because Christian elites have received modern education, they tend to exclude the participation of the community and traditional intelligentsia in the planning and implementation of environmental conservation projects. As a result, traditional intelligentsia and the community have been alienated. Related to the first is the modern education system, which tends to disregard indigenous knowledge. Traditional ecological knowledge is not included in the curriculum, while the teaching of local languages in schools only started recently. The third threat is the change in traditional livelihood practices and belief systems. These include change from traditional subsistence to commercial farming, change in staple food preferences, increased engagement in trade, hired labour and education, and conversion from traditional religion to Christianity. The final one is the weakening of traditional institutions that have been gradually replaced by the modern state and religious institutions. We suggest decentralising administration and ensuring participation of the community. Such a decentralisation process will build the capacity of Aari indigenous communities, improve Aari language education, and recuperate Aari traditions and practices. It will further develop, empower and support traditional institutions, and involve local communities in decision-making processes related to environmental conservation.

4. Conclusions

Overall, Aari's traditional religious beliefs, myths, and rituals assessed and discussed in this chapter suggest that the Aari's ways of living are guided by their worldviews. Although one of the two deities, upon which the foundation of beliefs is founded, is female, all *Babis*, *Babitoyədɔs*, and most *Godəmis* are men. Human/nature relations are, however, not necessarily gendered among the Aari. Because they value nature for its own sake and for its spiritual connotations, they seem closer to the eco-centrist view of the environment.

The natural environment is wisely managed. There are areas designated for use, and those reserved for rituals. The Aari also regard themselves as created from nature and their ancestral spirits and deities as incarnated in nature. There is a deep connection between the human and natural, and they are each other's keepers inter-existing in the environment. Ecological imperialism, with the formal educational system that disregards indigenous knowledge, the change of traditional livelihood practices and belief systems, as well as the deterioration of traditional institutions of environmental resource use and management are weakening Aari indigenous ecological thought, knowledge, and practices of conservation which are eco-friendly.

Aari community's traditional eco-friendly ontologies can be integrated with modern scientific knowledge and practices through community-led/based conservation efforts. The integration of traditional and modern practices can help create a more sustainable and indigenised communal conservation rooted in the communal beliefs of the Aari people. This should be done in an inclusive manner, without marginalising women, minority groups like the *mäna* (craftspeople), the Ethiopian Orthodox Church (which preserves lots of biodiversity around its churches throughout Aariland), the Qalä-həyəwätə Church (the protestant church with large following in Aariland), other Christian churches, Muslims, and other stakeholders. Such inclusive and carefully designed community-led conservation efforts that involve and empower their traditional keepers of the environment and create awareness through an environmental conservation discourse based both on a re-valuation of their traditional ecological knowledge and selective appropriation of modern ones will likely yield a more effective and sustainable environmental conservation in Aariland. Environmental conservation efforts in similar traditional societies elsewhere in Africa too would be effective if they follow this approach.

Acknowledgements

This work was supported by Arba Minch University under Grant number GOV/AMU/TH1/SORC/DELL/01/2010.

References

Akpomuvie, O. B. 2010. Culture and the challenge of development in Africa: Towards a hybridization of traditional and modern values. *African Research Review* 4:288–297.

Brooks, J. S., K. A. Waylen. and M. Borgerhoff Mulder. 2012. How national context, project design, and local community characteristics influence success in community-based conservation projects. *PNAS* 109:21265–21270.

Chilly, G. 2017. *Heritage, Conservation and Communities: Engagement, Participation and Capacity Building.* Abingdon: Routledge.

Chung-Tiam-Fook, T. A. 2011. Sustaining Indigenous lifeways through collaborative and community-led wildlife conservation in the North Rupununi, Guyana. PhD diss., York University.

Crosby, A. W. 1986. *Ecological Imperialism: The Biological Expansion of Europe, 900–1900*. Cambridge: Cambridge University Press.

Encyclopaedia Aethiopica. 2003. "Aari" by Carolyn Ford. Vol. 1. A–C. Wiesbaden, Germany: Harrasssowits Verlag.

Gyekye, K. 1997. *Tradition and Modernity: Philosophical Reflections on the African Experience*. Oxford: Oxford University Press.

Horwich, R., J. Lyon and A. Bose. 2011. What Belize can teach us about grassroots conservation. *Solutions* 2:51–58.

Huggan, G. and H. Tiffin. 2010. *Postcolonial Ecocriticism: Literature, Animals, Environment*. Abingdon: Routledge.

Ikuenobe, P. A. 2014. Traditional African environmental ethics and colonial legacy. *International Journal of Philosophy and Theology* 2:1–21.

Kaphagawani, N. D. and G. M. Jeanette. 2003. Epistemology and the tradition in Africa. In *The African Philosophy Reader*, ed. P. H. Coetzee and A. P. J. Roux. New York: Routledge.

Karbo, T. 2013. Religion and social cohesion in Ethiopia. *International Journal of Peace and Development Studies* 4:43–52.

Kebede, M. 2001. *From Marxist-Leninism to Ethnicity: The Sideslips of Ethiopian Elitism*. Paper presented at the International Conference on Contemporary Development Issues in Ethiopia, Michigan, 16–18 August.

Kebede, M. 2004. *Africa's Quest for a Philosophy of Decolonization*. New York: Rodopi Press.

Keller, J. E. 1988. *Revolutionary Ethiopia: From Empire to People's Republic*. Bloomington: Indiana University Press.

Kenenisa, F. 2010. The critique of modernity in postcolonial African situations. MA diss., Addis Ababa University.

Kidane, B., T. van Andel, L. J. G. van der Maesen and Z. Asfaw. 2014. Use and management of traditional medicinal plants by Maale and Aari ethnic communities of southern Ethiopia. *Journal of Ethnobiology and Ethnomedicine* 10:1–15.

Kidane, B., L. J. G. van der Maesen, Z. Asfaw, M. S. M. Sosef and T. van Andel. 2015. Wild and semi-wild leafy vegetables used by the Maale and Aari ethnic communities in Southern Ethiopia. *Genetic Resources and Crop Evolution* 62:221–234.

Kidane, G. 2015. Aari peoples verbal oral narratives' performance and social function. MA diss., Addis Ababa University.

Lepp, A. and S. Holland. 2006. A comparison of attitudes toward state-led conservation and community-based conservation in the village of Bigodi, Uganda. *Society and Natural Resources: An International Journal* 19:609–623.

McAlpin, M. 2007. Conservation and community-based development through ecotourism in the temperate rainforest of southern Chile. *Policy Science* 41:51–69.

Naty, A. 1994. From independent chiefdoms to Abyssinian subjects: The Aari interpretation of conquest and colonization. *Africa: Rivista trimestrale di studi e documentazione dell'Istituto italiano per l'Africa e l'Oriente* 49:498–515.

Naty, A. 2005. Protestant Christianity among the Aari people of southwest Ethiopia, 1950–1990. In *Ethiopia and the Missions: Historical and Anthropological Insights*, ed. V. Böll, S. Kaplan, A. Martínez d'Alo's-Moner and E. Sokolinskaia, 141–153. Münster: Lit.

Noguchi, M. 2013. Aging among the Aari in rural Southwestern Ethiopia: Livelihood and daily interactions of the 'Galta'. *African Study Monographs* Suppl. 46:135–154.

Oruka, H. O. 1991. Sagacity in African philosophy. In *Readings in African Philosophy*, ed. B. S. Oluwole. Lagos: Mass-tech Publishers.

Plumwood, V. 2003. Decolonizing relationships with nature. In *Decolonizing Nature: Strategies for Conservation in a Post-Colonial Era*, ed. William H. Adams and Martin Mulligan, 51–78. London: Earthscan.

Said, W. E. 1978. *Orientalism*. New York: Vintage.

Said, W. E. 1993. *Culture and Imperialism*. New York: Vintage Books.

Salvador, M. 2007. A modern African intellectual: gäbre-heywät baykädañ's quest for Ethiopia's sovereign modernity. *Africa* 62:560–579.

Serequeberhan, T. 1997. The critique of eurocentrism and the practice of African philosophy. In *Postcolonial African Philosophy: A Critical Reader*, ed. E. Eze. Cambridge: Blackwell Publishers.

Shigeta, M. 1990. Folk in-situ conservation of Ensete [ensete ventricosum]: Toward the interpretation of Indigenous agricultural science of the Ari, southwestern Ethiopia. *African Study Monographs* 10:93–107.

Soyinka, W. 1998. *Myth, Literature, and the African World*. New York: Cornell University Press.

SOZTD (South Omo Zone's Culture and Tourism Department). n.d. *General Overview of Aari People. 'yä 'äri bəherəsäbə 'ät'äqalayə gäs'əta'*. Unpublished Amharic language study report.

Teodros, K. (Ed.). 2001. *Explorations in African Political Thought*. New York: Routledge.

Tiffin, H. (Ed.). 2007. *Five Emus to the King of Siam: Environment and Empire*. Amsterdam: Rodopi.

Tolessa, K., E. Debela, S. Athanasiadou, A. Tolera, G. Ganga and J. G. M. Houdijk. 2013. Ethno-medical study of plants used for treatment of human and livestock ailments by traditional healers in south Omo, Southern Ethiopia. *Journal of Ethnobiology and Ethnomedicine* 9:1–15.

Turreira-García, N., H. Meilby, S. Brofeldt, D. Argyriou and I. Theilade. 2018. Who wants to save the forest? Characterizing community-led monitoring in Prey Lang, Cambodia. *Environmental Management* 61:1019–1030.

Verhelst, G. T. 1990. *No Life without Roots: Culture and Development*. Trans. Bob Cumming. London: Zed Books.

wa Thiong'o, N. 2013. Tongue and pen: A challenge to philosophers from Africa. *Journal of African Cultural Studies* 25:158–163.

Weis, T. 2015. Vanguard capitalism: Party, state, and market in the EPRDF's Ethiopia. PhD diss., University of Oxford.

Yntiso, G. 1988. Traditional labour organization of Aari. BA diss., Addis Ababa University.

Yntiso, G. 1995. The Ari of southwestern Ethiopia: An exploratory study of production practices. *Social Anthropology Dissertation Series*, no. 2. Addis Ababa: Addis Ababa University.

Yntiso, G. 1996. Economic and socio-cultural significance of enset among the Aari of Southwest Ethiopia. In *Enset-based Sustainable Agriculture in Ethiopia*, ed. S. Brandt and G. Seifu, 107–120. Addis Ababa: Institute of Agricultural Research.

Yntiso, G. 2010. Cultural contact and change in naming practices among the Aari of southwest Ethiopia. *Journal of African Cultural Studies* 22:183–194.

Yntiso, G. 2015. On Aari people's culture and history. *'ä 'əri nəq'ašə* 3:43–52. (South Aari district's yearly magazine in Amharic and Aari languages).

9 Climate Injustice

How It Is Affecting Africa

D. A. Masolo

It is April in the year 1964. Kenya had just become independent from Britain on December 12 of the previous year. And then Jaganyi son of Wanyande (Jaganyi ka Wanyande) dies at the age of about 90. He was the more famous older brother of my maternal grandfather Senior Chief Oloo Wanyande of the Basonga clan. The people of Usonga could not have cared less about the frequent colonial orders that were communicated to them by the government's administrative representatives like their own Chief Oloo. They cared more about the concerns that Jaganyi helped them address and resolve because these concerns were about their livelihoods. Together with Nyahasoho, one of his many wives, Jaganyi had endeared himself, and his large Wanyande family, to the people of Usonga and beyond. He was a rainmaker, and Nyahasoho was his ever-present apprentice. She fetched the herbs and roots that were boiled together in a large earthen pot in the darkest belly of her round house. Together, they were the climate keepers of their people. At his funeral, people reminisced how many times in their memorable past he had saved them from an impending hunger and famine due to a devastating drought. The latest famine of 1941–1942 caused by the devastating failure of the long rains overlapping the two years. They baptised this famine "Osembo," a name coined from the verb *sembo*, to escort, to gently assist to walk or push gently. Because of the severe drought, pastures had dried up, and many cattle keepers had been forced to "escort" their cattle to be temporarily kept by relatives who were in relatively better locations. Because the cattle were emaciated and weak, they had to be handled gently as they walked the long distances to places of rescue. This scene happens repeatedly in many pastoral regions of Africa. In the linguistic region that included the then drought-stricken Usonga, people mark unusual events in their traditional calendars – such as the "Osembo famine in this case – by giving children born within those times the same name as the one they coin for the event. That way people can tell people who belong to the same age set, or how old they probably were relative to the event in question. I came to know many people from both sides of my ancestry who were called by this same name. So, when the Roman calendar came, we could easily tell that they were born in the 1941–1942-famine season. Most communities

DOI: 10.4324/9781003287933-9

have their own indigenous calendars, and cross-linguistic comparisons help in the determination of shared experiences as well as different ethnic markers or designations of common historical times. There must have been other, similar markers of this cross-ethnic famine from adjacent communities in the region. Because of this practice of naming children after big events, one comes to know the ages of these individuals by historical association. Thus, among the Luo, and possibly in other communities as well, there are many individuals who bare names they were given to commemorate and honour historical figures like, among the most popular, Julius Nyerere, Patrice Lumumba, Milton Obote, Nelson Mandela, Winnie Mandela, Jomo Kenyatta, Ahmed Ben Bella, Hastings Kamuzu Banda and, most recently, Graca Machel, Barrack Obama and Michelle Obama. Such individuals become living markers of their times. To name persons after a devastating event like hunger and famine is an indicator that the interest in the practice is the recording of history rather than the glorification of persons. The exception is the bias against elevating people of questionable character to such social honour, a widely recognised stand around the world. Recently as the world came to its moral senses, statues formerly honouring racist individuals around the world have been pulled down to minimise the visible memories of hate and oppression and spurred mass movements like Rhodes Must Fall and Black Lives Matter.

As lauded in the mourners' adages and in subsequent lore about his service to his people, Jaganyi together with his wife Nyahasoho were credited with stopping the locust invasion of 1941–1942 after they were reported to have been seen headed south-south-west and over the Samia hills toward Usonga from across the border in eastern Uganda. A well-known herbalist and fierce opponent of the surrogate rule of Mumia (*Loch Ka-Wango*) on behalf of the authoritarian colonial rule, Jaganyi had previously endeared himself to his people through his ability to cure such new, epidemic-level diseases like leprosy and smallpox. Also, he already had been known to be able to control a variety of symptoms of mental health problems generally. Basonga people's perception of Jaganyi was that he was their medicine-man in healthcare and hero against impending climate-related calamities. These abilities were often related to his heaves to distant places beyond Usonga and sometimes even beyond the known towns of Kenya. When people heard that he travelled to then-Tanganyika, they related his medicinal knowledge to the mythical prowess and legends about the traditional medicines of Tanganyikans. But when he visited his two sons, the first one, Lucas Kaka, on the island in Lake Jipe in the Taveta region of Kenya, and then Alex Ongode in Korogwe, a small town in the dry region in the shadow of Mt. Kilimanjaro that lies between Moshi and the coastal town of Tanga in 1951, it had not rained for a while, and the drought had brought panic and worries to the people of the town and its surroundings about an impending hunger and famine. His gentle and calming voice in greetings to the people he met, saying, "No need for panic, it will rain," appeared not to allay the fears of the people of his host town. But, a few days later, it rained in Korogwe, and it rained in torrents. Legend has it that it rained in the entire region that runs to the Lake Jipe area where Kaka lived. News about a "medicine-man from Kenya" spread quickly like wildfire, and they thought he had "brought" the rains. So, while his legendary medicinal abilities were sometimes attributed by clients and the people of Busonga and surroundings generally to his learning and importations from Tanganyika, the people of Lake Jipe and Korogwe, the people of Tanganyika themselves thought he had powers from his native Busonga in Kenya. Jaganyi's fame had come full circle. According to his own family, however, Jaganyi owed all these attributes of power over nature to his brother-in-law Odenga, Nyahasoho's brother from Msware, a village of the Samia clan extended across the border in Uganda.

Back at home in Busonga, large delegations would come frequently to Jaganyi's home in the swampy lowland adjacent to Lake Victoria in large delegations each time the Basonga sensed a looming climatic crisis. They transformed Jaganyi's home into their version of a meteorological or weather-and-climate centre. While visits for healthcare needs were usually private, visits about climate crisis were public and loud. They came to his home on these occasions to sing his praises as they urged him to perform his miracles to save them from looming crises, and then, in apparent consensus, to take action to mitigate what they understood from the past to be the antecedent events that were, in their memory, often followed by devastating experiences of hunger and famine. If the rains were delayed or looked like they would totally fail, and if reports of spotted locusts or army warms gave them a sense of collective threat, Jaganyi was there to perform his miracles. They believed that powerful medicines from Tanganyika – but really from his in-laws across the close-by border in Uganda – would prevail over nature. Jaganyi would make his concoctions to ward off any danger to his people. After all, he had played an important role in the rebellion of the Basonga, formerly under the traditional leadership of his father Wanyande k'Ohang'o, against what they viewed as corrupt and unwarranted subjection of the Basonga to unwanted domination. Busonga was, like the adjacent Bunyala, then a Sub-Location of Samia Location under Kadima Shiundu, a brother of Mumia Shiundu, the infamously dictatorial surrogate of the colonial establishment in western Kenya. Kadima ruled over his Location from 1911 to 1927. He handed power over to Mukudi Nyamwandha who then led Samia from 1927 to 1948. His successor, Kanoti Okwaro, who led the Samia Location from 1948 to 1952, oversaw a huge loss of the Mumia power in the region when Busonga became a full-fledged Location under its own Chief in 1952. Oloo Wanyande, Jaganyi's half-brother who had been Assistant Chief from 1941, became the first Chief of Busonga. He became Senior Chief in 1959 and served in a supervisory role over the Location until he finally retired after independence, and not before overseeing another crucial campaign for their people – this time by resisting the transfer of Busonga to Western Province, a move they saw as an attempt to make the Basonga a Bantu rather than a

Luo community. After leading a delegation to Kenyatta in 1963, they were granted their wish and Busonga was annexed to Alego in new parliamentary boundaries as Alego-Usonga Constituency.

Clearly, the role of Mumia's power and the resistance to it by some of his own subjects and in surrounding communities need to be understood in the context of the two colonial-owned World Wars (1914–1918 and 1939–1945). The forceful drafting and shipment of natives to go fight in the wars in distant places on behalf of the Whiteman was a violation of the freedoms and other human rights of natives. In addition, the confiscation of property, especially cattle and grains to feed soldiers on the battlefields as well as the levying of taxes to finance the wars and other colonial projects were like insults on top of injury. In every way, these colonial actions were unwarranted and unacceptable. Whatever and regardless of other consequences of their resistance against the institutions of colonial rule, rebellious leaders like Oloo may have become popular to their people, but they also became losers in regard to some of the benefits that submissive local leaders or colonial Chiefs and Church leaders reaped either directly as government servants or by sending their children to colonial and missionary schools. Across the British colonial world, this is how the first generation of post-independence leaders in all sectors was minted. Rebels like Oloo Wanyande, on the other hand, descended into an abyss of illiteracy, invisibility, and poverty. Here too, Jaganyi became an enigma. His few children who had secured jobs in the Railways network built their own social network and brought their other relatives to live and go to school where they had been posted in East Africa, especially in Tanganyika where they finally found employment and further built their own paths to high positions in society. Today, one of his grandsons, Deogratius Jaganyi, is Vice-Chancellor of Mount Kenya University. His father, Alex Ongode Jaganyi, served in the Tanganyikan police service where he rose to the rank of a regional Police boss in Ukerewe, an island in Lake Victoria where he had moved from Korogwe. All these individuals are directly related to or descended from the line of the rainmaking family. Nyahasoho's brother Odenga from Msware, who purportedly either taught Jaganyi the art of rainmaking or "passed the powers" over to him, was Deogratius's maternal grandfather. Several of Nyahasoho's other descendants have also made their way into high social positions in Kenya and abroad. Jaganyi's rainmaking has left a visible trail of good social climate to many of his descendants.

The moral of the story is that either through his own actions or through roles in collective actions with others in his famous family, Jaganyi was accustomed to visibility in the eyes of his Basonga clan. In battles for territorial claims against their Bantu-speaking neighbours, their victories were viewed as attributable to his prowess as a medicine man and spiritual leader of his people and psychologist of their young warriors. He led his people and often scored victories for them against adversaries, both human and natural. According to Titus Jaganyi (aka Taro), Alex Ongode's nephew from his sister Hamtal, the rainmaking ritual was a meticulous process defined by a series of events that started before sunrise. A young, pre-puberty girl would be sent to fetch water from the river under the strict warning not to seek help with placing the pot of water on her head or speak to anyone along the way once the pot was on her head. Jaganyi would boil his concoction of herbs in this water away from the prying eyes of other members of the home or village. The concoction would boil all day, or for as long as it would take to "make rain." The same young lad who had fetched the water in the morning would fetch some more and would be the only person allowed to move the lid from over the boiling pot to look inside. If she "saw clouds" and "heard the rumbling of thunder" coming from down the big pot, "rain was ready," and it was said that at the very same time, it would rain in Busonga and its surroundings. And rejoicing would commence, perhaps a bull would be slaughtered as an offering. Like with most practices of exclusive expertise, the herbs were secretive. Titus (Taro) appeared to be the chosen heir to his grandfather's secrets, but the grand Jaganyi died in April 1964 while Taro was still away in Tanganyika.

Time, Space, and Environmental Control

Since Jaganyi's time, space and meteorological sciences have grown. Scientists understand better now what the patterns of repeated natural occurrences like the annual migrations of marauding locusts are. Space science with satellite pictures of the earth can now tell us in real-time where exactly locusts are breeding, and when they are likely to start their migrations. Yet this knowledge does not necessarily replace the specific indigenous knowledge of local peoples everywhere around the globe.

The people knew how to read climate signs. They could observe keenly the gradual disappearance from the Eastern skies of the early evening star seen clearly those days in late February and the first part of March. Its disappearance and replacement with the south-south-eastern clouds during this period was the sign that the rain-bearing Monsoons were moving into the region so preparations of the fields in readiness for planting would be kicked into faster gear. They knew not only how to tell barren clouds from the rain-swollen ones, but they could also tell quickly what unusual sun blockers were. Then they would send emissaries to distant neighbouring places to bring back any climate-related or pest-related information. Sharing this kind of logistical information over regions advanced food security and reduced the likelihood of dependency. Even today, people still read their centuries-old climatic networks and time their activities accordingly. When they find themselves off-target, they adjust as frequently as may be needed. Scientific knowledge based on satellite maps of the earth can lead to the prevention or lessening of the spread and impact of locust invasions through well-timed aerial spraying, but local people know well the natural cycles or climatic patterns and the likely re-emergence of the earthly creatures that follow these patterns.

Jaganyi would not have been of much help to his Basonga people if the locust swarms of their time had been of present-day proportions. It is also unlikely that he would have had the required knowledge, skills, or tools to achieve what his Basonga people had come to expect him to be able to do for them. But these premises are themselves unreliable for the following reasons: they are hypothetical, and they also assume that much of what counts as knowledge is evidence-based. As shown by Barry Hallen and J. O. Sodipo (1997) about the Yoruba distinction between knowledge and belief, if we relied on individual evidentiary criteria for knowledge, then most people would not make much knowledge claims as much of what most people claim to know they have come to believe because they trust the chain of the testimony through which their beliefs come about. Thus, Jaganyi's practice hangs on the testimony of his people, and the stories they told other people about it. This is social epistemology, which is grounded in the claims of other, socially trusted people. Thus, the people of Jipe, Korogwe or Ukerewe depended on the legend that preceded the man, and their beliefs gave him the responsibilities, and they the expectations that either side could not run away from.

Although the world was in the middle of a World War in 1941–1942, the human and material costs of the war could be ignored by millions of rural African people as only relatively small numbers of their fellow countrymen were drafted or conscripted to fight in the colonial armies. The climatic threats to collective survival were far greater. Also, the earth's temperatures may already have been rising unnoticed, although they always have been, but the effects could be absorbed by small adjustments that people made because failures were neither overwhelming nor were inexorably unstoppable. By proportion, a protracted drought and related locust invasions were likely to be followed by plentiful rains and even bumper harvests in the following seasons. But more importantly, social networks were strong and people voluntarily transported gifts to affected neighbours and relatives. Inter-community marriages and resulting social networks and collaboration made it possible for people to seek and receive assistance across community borders in times of need. The sharing of the values of good times and the suffering of adversities in bad ones brought lessons of equality and peace across regions.

Jaganyi shot to fame among his people in relation to what they believed were his climate actions in the locust invasion of 1941–1942, and the consequent "Osembo" hunger and famine in the same span of time. Colonial Kenya had suffered Smallpox outbreaks in 1934 and 1943, which years coincided with the timing in some of the narratives about Jaganyi's ability to heal what they called "new diseases." But, as explained above, his fame really came from his perceived role as the "Climate Man" for his people of Basonga, a view which exemplifies one of the central issues in social epistemology, namely the role of popular testimony – in this and similar cases, the biographical narratives that often travel way ahead of their subjects.

Narratives about Jaganyi's healing and rain-making powers had travelled far and wide, perhaps through friends and relatives who were traveling through what was then a free-travel zone of the colonial East Africa and straddling the first decade and a half of the independence period.

The African variety of locusts is known as "the desert locusts," and it breeds in moist sandy or sand/clay soil where they lay their eggs about 10–15 cm below the surface. Unlike the Sachedas of north America which go into hibernation for 17 years following the last year of their emergence, the desert locust can emerge every year when the conditions are favourable, but they do not follow any known, consistent breeding pattern, suggesting that although their emergence may be unpredictable, the degree of their devastation may, however, be dependent on the presence of favourable conditions such as high atmospheric temperatures. Also, in contrast to the north American Sachedas, the desert locust is migratory. And this is what has been making the African desert locust to be more deadly and more widespread in the past several decades due to the gradual atmospheric warming and climate change resulting from the greenhouse emissions of an increasingly industrialised world. In the earlier times, the desert locust has been known to breed in smaller regions and their spread has largely been naturally contained or limited, and their effect minimised. The last time the desert locust caused a major plague was in 1986–1989, resulting in the great famine of the Horn of Africa, also known as the Ethiopian famine. In the pre-industrial boom period from which Jaganyi emerged as a hero the locust invasion was therefore not only a rare occurrence, but the geographical coverage of its damage may also not have been terribly great, and its effect may have receded relatively quickly as the regional climate conditions readjusted to their normal cycles. Many other droughts have occurred since 1941 and 1942, and people have experienced multiple episodes of hunger and famine since then. The 1980–1982 hunger and famine in East Africa were major. Jaganyi's people, and the entire Luo community christened it the "Gorogoro," named thus for the 2Kg. – tin butter container that subsequently became a standard market tool for measuring grain volume. And the world-known 1983–1985 famine in Ethiopia due to drought was declared a world disaster that claimed anywhere between 200,000 and 1,200,000 lives, and worsened by the locust plague of 1986–1989, thus frustrating or delaying any recovery in the region. Social networks – local, regional and global – were like in Jaganyi's Basonga and in many African communities across the continent, the basis of the mitigation of suffering resulting from them.

Climate change has expanded the breeding regions for the desert locust, suggesting that the locust can adapt to new conditions, and thus expand the space of destruction beyond previously imaginable boundaries. Even as we write now, there is a major, multi-continent looming hunger and famine that will affect far greater populations than ever seen before due to impending, monstrous locust invasion unless the hatchlings are decimated before they grow wings. While the causes of the 1983–1985 famine in Ethiopia are many and complex, including bad government agricultural policies under the dictatorship of the then-president Mengistu Haile Mariam, the causes and management of drought and pest invasions can equally come from bad, politically pushed economic policies. The current, monstrous locust threat is no exception. Its monstrosity is due in part, as already seen in Kenya, to the fact that they will for the first time invade even areas of high elevation with what used to be cool climatic conditions where they traditionally have not been known to farmers. Why is this so?

Locusts are dry-region, warm-temperature pests, so the recent expansion of their breeding boundary suggests that the favourable conditions of breeding have spread beyond the formerly known traditional grounds. The scientific explanation is this: locusts will not abandon their culture, at least not within a short time from an evolutionary standpoint. But they can travel or spread as far as the conditions of their survival and thriving allow. East Africa has experienced easily noticeable climatic changes that are devastating local economies and disrupting people's cultural activities. Once dry areas where populations had for centuries practiced goat rearing, parts of the North-Rift region were devastated by constant torrential rains that disrupted scholastic activities in addition to arresting their economic activities still. Schools and homes were swept away by the rains as many local people, not used to traversing waters in their daily movements, were caught unawares and drowned in flash floods. Water and boats were never part of their lived lexicon. But suddenly, their very survival is now sometimes made to be dependent on how to use boats to navigate life through the newly formed rivers. In other words, global climate change affects everyone. Thus, the already vulnerable peasants suffer the effects of climate change the most. Here is the problem, at least from a conceptual standpoint. For decades, and among the innumerable strategies offered to African nations by donor nations of the global North, sustainability has been one of the key concepts, and it often seems to make sense. It is also a term that counsels against wrong perceptions of development to be grounded in the infrastructural use of space. In this sense, the word sustainably refers to a concept that works especially well for defining change in terms of what can be supported or managed with locally available resources so the achieved change can have long-term positive and locally repeatable and manageable effect. But current unpredictable, meaning unreliable climatic conditions are making the idea of development meaningless as peasants are forced to abandon their centuries-old activities or in order to try new productive practices, require long-term harnessing before they can become equally unsustainable.

That global temperatures are rising is no longer an issue of debate. Images of melting ice in the North Pole have been seen everywhere and, in the tropics, once perennial rivers have dried up, leaving fruit-producing regions in high elevations scrambling for alternative agricultural occupations. Peasant farmers once considered resilient suppliers of produce to their local markets now must travel distances to purchase what they once supplied to cities and other parts of the country. Mounts Kilimanjaro and Kenya have lost much of their fabled snow that once gave them fame as tourist attractions. Nairobi, sitting at 5,500 feet above sea level, is now 1.8°C warmer than it was in the 1970s. In ecology, formerly cooler regions once regarded as lying outside the "ecological boundaries" of the locust cycles are now becoming hospitable to the pests. In addition to the locusts, other, never-before-seen pests have emerged. The "Army warm," now experienced annually for the past 15 years, has also made its footing as an additional plague to be dealt with. While locusts devour any and every vegetation in its path, the "Army warm" embeds itself in the "heart" of the 1–2-foot stalks and cuts life out of them. Being invisible to the casual observer, its destructive presence is often noticed too late when pesticide spraying no longer helps. In poor communities across the African continent, and perhaps elsewhere around the globe too, climate change is displacing people, disrupting lives, and thus spreading or deepening poverty and dependency.

What, then, is warming up the earth's atmosphere? Scientific figures are disguisingly misleading. Agreed that the amount of CO_2 in the atmosphere is responsible for global warming, the community of sceptical scientists argues, again with the concurrence that CO_2 makes up only 0.04% of the gases in the earth's atmosphere, and that of this, only 0.0016% is human-made while the rest, about 95%, comes from natural sources that include volcanos or decomposition processes in nature that the outcry about climate change as a result of global warming is either exaggerated or outright false. If things remained as intact as these statistics suggest, in the best-case scenario, my earth's atmosphere which I encounter daily would probably be at least closely similar to the atmospheric state that I experienced about 70 years ago. But something more seems to have happened that did not happen 70 years ago, besides the obvious increase in the human population.

Over the past 70 years, atmospheric CO_2 has been on the rise. With the rise of the human population, improvements in scientific and technological knowledge have led to increase in industrial growth. Unpredictability of effects from manufacturing and from the use of the products themselves can be correlated to the increase of CO_2 in the atmosphere. The peak of industrial growth in the West in the 1960s and 1970s led to greater demands for fuel. As a result, oil production and sale became a source of great political and economic power for countries that had discovered large oil deposits within their territories. The formation of OPEC in Baghdad in 1960 (founding members: Iran, Iraq, Kuwait, Saudi Arabia, and Venezuela, and later joined by Indonesia, Libya, United Arab Emirates, Algeria, Nigeria, Ecuador, Gabon, Angola, Equatorial Guinea, and Congo; these latter two joining the group as recently as 2017 and 2018, respectively) gave birth to an alternative point of attention in the otherwise dual geopolitical division of the world into East-West blocs with willing allies on both sides. In direct relation to this oil boom, oil refining expanded, and tanker shipbuilding and sea commerce in the transportation of crude oils around the world became lucrative business but also marked the degradation of the seas due to pollution. At the same time, landlocked countries increased their dependency on road transportation of refined petroleum, thus resulting in the emission of CO_2 directly into the atmosphere from surface-based oil tankers, which travel hundreds and sometimes thousands of miles to transport the commodity.

Not only did the formation of OPEC give member countries a huge political leverage, but these countries now had access to huge financial resources that could be invested in modernisation as well as consumer goods. Industries, especially in the global north, expanded to feed the demand for manufactured goods at home and abroad. Not only do these factories emit CO_2 directly into the air, but large amounts of the gas are also produced when cars and power plants burn coal, oil, and gasoline. Many countries, not limited to oil producers or its biggest consumers, saw economic boom at the expense of quality air and general environmental well-being.

Because of the previous statistics regarding gas components in the atmosphere, sceptics of the degree and real effect of human-caused CO_2 in global warming have argued that the human factor in global warming is negligibly minimal compared to other, natural causes. And because the Paris Accord requires agreement to the reduction of greenhouse-gas-emission, ostensibly reducing or drastically transforming how much CO_2 is pumped into the atmosphere, particularly by coal-burning power plants. Industrial countries, especially those dependent on coal for generation of power, have mounted opposition to the Paris Accord. The world's projections are that if the developed world could do their part so the global CO_2 emission is reduced by 25%, the world could reduce the earth's warmth to 2° above pre-industrial levels by 2030; and if the reduction is by 55%, the global ideal of 1.5° above pre-industrial levels by 2030 could be possible. As you can see, significant sacrifices are required for us all, the global community, to do something in order to save the earth. These were the projections based on the Paris Accord in 2010.

In all this, Africa's CO_2 emissions for the entire continent were negligible, producing a meagre 2–3% of the world's volume, and of this, South Africa, due to its heavy dependency on coal for power production, accounted for 1% by itself. The economic scenario is obvious. South Africa's level of industrialisation outmatches that of the rest of the continent put together. Nuclear wastes and resulting environmental destruction – not just degradation – from some of these (now de-commissioned) power plants is a problem that stares South Africa down as the country contemplates its safer future. Intentionally kept uninformed by the apartheid regimes, South African black communities who were pushed to live near these plants are now crying out for help out of the once-unspoken crimes of the white apartheid rule. Here is my hypothetical argument that makes the crucial point directly related to the idea of climatic injustice:

> If global warming is due in significant part to emissions of CO_2 into the atmosphere from industrial activities, and that the effects of elevated levels of CO_2 in the atmosphere affect everyone globally and indiscriminately, and if it is scientifically proven, as it has, that reduction of human-caused emissions of CO_2 at least lessens or slows down the current state of global warming, then those countries with the greatest number of industries that emit the most CO_2 into the atmosphere have the highest

responsibility to reduce their portion of this emission as agreed at all the Protocols of the World's Climate Change Conferences. Refusal to do so in any form does injustice both generally but especially to those countries or populations of the world who are least culpable for the pollution yet most negatively impacted by the effects of global warming.

As shown in the compliance/performance assessment of the decade at the World Climate Change Conference in Madrid in December 2019, there was a gap between what was pledged and what needed to be done, so the world has virtually failed vis-a-vis the 2010 target agreements. Only one region, the European Union, lowered its emissions; another country, India, slightly lowered its emissions, while the United States, Russia, and China showed no change, and Brazil and Indonesia had even higher emissions.

Talking about 2019, one is reminded that Donald Trump was in-charge as president of the United States and, playing on the broadened view of his rhetorical phrase, "America First," he whipped up the Republican sentiments in opposing the Paris Accord because "[it] was very expensive, unfair, job-killing, and income-killing," which, to the Republicans-turned-ultra-conservatives under Trump, ran contrary to their ideals of individual freedom, limited government and free market. Joining the United States in opposing the Paris Accord were Brazil, Australia and Saudi Arabia. Trump eventually declared the United States' pull out of the Accord altogether.

So let me go back to the Jaganyi factor in my backyard. The whole purpose of the Basonga people's gathering at his home in the face of impending calamities of "climate out of pattern" – neither drought nor locust invasion – was their sentiment that matters beyond individual action required communal action. His action was a form of representative agency for collective action; he would perform on behalf of the community of which he was part. His leadership and role bequeathed upon him by his people put on him a responsibility he could not renege, and their collective aim was to make the environment useful to all. It was the source of their livelihood. Hence, their laudatory song for him:

> Jaganyi, you rescued us from an impending calamity
> Because of you we averted sure destruction
> Of our environment and us
> You son of Wanyande
> You brother of Oloo
>
> Jaganyi you saved us from a sure locust invasion
> As they already covered the sun
> As our destruction looked inevitable
> You brother of Otete
> You made them change their route;
> After they had destroyed Ugenya and Samia across the river
> And when they were headed for us, you made them drown in the river

> And you saved us
> You son of Wanyande
> You brother of Oloo
> You brother of Otete.
> And when we faced drought,
> You brought the rains
> And it rained until there were floods
> And you tamed the torrents
> So we got our harvest

What would have happened if Jaganyi had done nothing about the matters to which his people had drawn his attention? There are many Jaganyis across the continent, people whose alleged "powers" are claimed in their people's folklores and narratives. Appeals to them are not as important in scientific terms as what they symbolise or epitomise: that human survival depends on certain dependable or predictable global patterns, such as acceptable amounts of the important greenhouse gas, or CO_2. In addition to this fact, such appeals also symbolise the awareness of the long-predictable seasonal climate rhythms which they have come to take for granted as part of nature – just like they take for granted that the sun will "rise" every morning. When these basic assumptions of life are suddenly interrupted, panic ensues because livelihoods suddenly appear threatened. So, the issue is not that village elders are suddenly regulators of atmospheric gases on which life depends, but the awareness that when things seem to fall apart, there is someone . . . in that long chain of responsibility, from village elders to government administrators and political authorities, who alone can influence and regulate the policies under which acceptable amounts of CO_2 are discharged into the atmosphere. So, the song designed previously for Jaganyi could well be posted to the good leaders under whose leadership governments have adhered to such regulatory agreements like the Paris Accord.

It is about accepting common responsibility to do the right thing because we are mutually responsible for each other. Western governments and the industrial establishments that operate under their protection do not believe we are capable of industrial knowledge and production, or that we need it, and least of all, that we deserve either to know or to make noise about the adverse effects on us of Western industrial power/empire. For centuries, the Western world, and now China, believe that our existence has no consequence for the future of the world, except when they believed that Africans will likely wipe them out with Ebola or HIV/AIDS, but not when they are about to wipe us out with Covid-19. Hence, they can manufacture vaccines that will be impossible for most African countries and the developing world, in general, to safely store and effectively use. Thus, the world continues to operate under these conditions of epistemic injustice and drives the morals and ethics of the production and distribution of human needs for properly defined global survival.

Framing mutual moral responsibility for the production and sustenance of the conditions under which people can live without threat to others is what has been called by the term "Ubuntu." Because it has now been diluted and in some cases even drowned under misuses and abuses in some recent studies (see, for example, Shutte 2001), it is prudent to keep off a long analysis or interpretation of the term here. James Ogude's trilogy (Ogude 2018, 2019; Ogude and Dyer 2019) on the multi-disciplinary approach to understanding the concept as a socio-political critique of disruptions of human ideals remain the best and most widely discussed to this point. First, let us say that the term (Ubuntu) as popularised by the now-widely quoted translation by The Rt. Rev. Desmund Tutu (2011), *Umuntu ngumuntu ngabantu* (roughly: persons become persons through other persons) carried a crucial ethical principle that is often drowned in explanations that see it only in political terms. As magisterially explained by James Ogude (see "Introduction" in Ogude and Dyer 2019), the term emerges as a critique of socio-political orders that stand or draw their strength by appealing to privileged sections of the society while defining everyone regardless of the contradiction, as both dispensable yet also irreplaceable support of their economies. In his use of the term, and in his English translation, Rev. Tutu meant to decry the fragmentation of society while lauding, by the same stroke, the recognition and moral superiority of mutual dependency among humans. In this and other senses articulated in Ogude's trilogy, the term "Ubuntu" is a strong critique of narcissism and other ethical and socio-political norms, which build themselves on discrimination of others.

American rejection of and brief exit from the Paris Accord under President Trump was built upon focus on the destructive idea of individual freedom, wealth accumulation, and free markets, the kind of moral that was once referred to as "the principle of *Lazier faire*," let everyone practice their unhinged freedoms – the cardinal values and pursuit of conservative politics. In that mindset, any principle or goal that falls short of the conservative ideal is branded "socialist." But what is wrong with socialism anyway? Under the principle of "Ubuntu," human actions are considered morally good if they are not intentionally designed to cause harm to others. The South African apartheid did not cause the invention of the term "Ubuntu" in the vernaculars of the so-called Nguni group of languages or dialects in central and southern Africa. Instead, it only presented precisely the kind of social, political, economic, and moral orders that obviously were not built on the kind of ideals that are captured by the term. In Tutu's theological rendition, God was not a Christian, but one becomes a Christian when they believe in and practice the ethical ideals taught and exemplified by Christ. There seems to be a rift, then, in his view, when one calls themselves Christian and yet believes in and practices the kind of discriminatory principles like those of the apartheid Whites of South Africa did, or like those of the American Evangelicals when they call themselves Christian yet believe firmly in the kind of world regulated by racism and other forms of discrimination.

Note

1 Francis Bacon Quotes. BrainyQuote.com, BrainyMedia Inc, 2021. www.brainyquote.com/quotes/francis_bacon_100764 (accessed December 2, 2021)

References

Hallen, B. and J. O. Sodipo. 1997. *Knowledge, Belief, and Witchcraft: Analytic Experiments in African Philosophy*. Revised edition. Stanford: Stanford University Press.
Ogude, J. 2018. *Ubuntu and Personhood*. Trenton, NJ: Africa World Press.
Ogude, J. 2019. *Ubuntu and the Reconstitution of Community*. Bloomington: Indiana University Press.
Ogude, J. and U. Dyer. 2019. *Ubuntu and the Everyday*. Trenton, NJ: Africa World Press.
Shutte, A. 2001. *Ubuntu: An Ethic for a New South Africa*. Dorpspruit: Cluster Publications
Tutu, D. 2011. *God Is Not a Christian: Speaking Truth in Times of Crisis*. London: Rider.

10 International Environment Law, the Humanities Nexus, and Some Reflections on *"Creative Legal Solutions"*

George Odera Outa[1]

1.0 Introduction: In Search of Creative Legal Solutions

The aim of this chapter is to contribute to discussions around the efficacy of some of the well-established global mechanisms for sustainable forest management, with a view to advancing a regime of *"creative legal solutions"* for forest protection. Such discussions and solutions, it is argued, are somehow missing in present debates. It is further argued that a discussion on global mechanisms for sustainable development should trigger ideas and solutions, while also enhancing the quest for deterrence against exploitative misuse of the bounties of mother nature.

In demonstrating the apt relevance of the humanities, the chapter draws comparative insights from the world of creative literatures; specifically, Chinua Achebe's well-known novel: *Things Fall Apart* (1958/1966, referenced hereafter, as TFA); with a view to learning from the signature creative portrait of a precolonial African community's handling of the complexities associated with significant environmental extremes; be they cultural or more directly, climatic. The solutions require conceptualisation in terms embedded in culture and indigenous knowledges that African societies offer, especially on how to live in harmony with nature.

1.1 Structure and Outline

The chapter begins by first outlining some of the main standards and instruments that are applicable to forest protection; including the international conventions that govern environmental protection. Thereafter, we discuss the "state of play" around the contemporary "tragedy" associated with unending forest exploitation and decline. The chapter then proceeds to analyse *two* legal approaches that are ostensibly effective in providing some form of deterrence although, on closer scrutiny, this might not necessarily be the case. The chapter suggests four alternative approaches that can enhance forestry protection and additionally attempt to re-draft more stringent clauses for the existing legal regimes. Whereas the alternative approaches we suggest are not necessarily new, our aim is to present creative ways in which to

DOI: 10.4324/9781003287933-10

read, strengthen and possibly derive more from existing approaches. In the final analysis, we reflect on the additional lessons offered through creative literatures, and in particular by way of the referenced Chinua Achebe's text already mentioned; not so much because this was Achebe's main mission, but rather to point to a less-researched pathway: creative literatures as a source of divinity, wisdom and knowledges, including persuasive legal possibilities.

2.0 The International Environment Protection Regime

International law lacks an effective legislative body, a central leader (with associated administrative apparatus), an overarching judicial system with universally recognized interpretive and enforcement power, and a central system of compliance and punishment.

(Takacs 2016: 1)

***

The demand for the conservation and preservation of the environment for the sake of "present and future generations"[2] is today edged in a steadily growing number of multi-lateral and bilateral agreements. They are also to be found in the corpus of customary law norms and principles that can be gleaned from the growing discipline of International Environmental Law (IEL), as well as in specific member state actions. Member State actions consist primarily of the various pieces of domestic legislation that have been enacted in the attempt to domesticate the key principles and imperatives emanating from those self-same international agreements and conventions. From an International Law standpoint, the relatively recent "Paris Agreement" (2015) would, therefore, be simply "soft law" and as it were, one of the protocols of the *UN Framework Convention on Climate Change* (UNFCCC 1992) that today, provides the overriding global framework for dealing with the advancing reality of climate change.[3] However, ever since the Stockholm Declaration (1972) there has been a steady growth of agreements and conventions, with the United Nations, *Non-Legally Binding Instrument on All Types of Forests* (NLBA 2007) being one of the more recent ones in so far as a unifying global forest protection regime may be concerned.[4] Given the Non Legally Binding Agreement's (NLBA) singular centrality in this chapter, we further examine its import and attendant challenges in subsequent sections.

In the meantime, one source has put the number of IEL-related multi-lateral agreements at close to 2000, if not more[5] and that there are some notable ones such as the *Convention on International Trade on Endangered Species* (CITES 1973); the *Vienna Convention on the Protection of the Ozone layer* (1985), *Convention on Biological Diversity* (CBD 1992), among others, whose import and significance are clearly outside the scope of our present

International Environment Law, the Humanities Nexus 155

focus. The much-established International Law principle of "present and future generations" does, of course, find expression in other terminology as "*inter-generational and intra-generational equity*" which has quite, substantially, been accepted as "cannons" of environmental protection. The import of such principles as "international cooperation" is in recognition of the fact that certain matters go beyond state sovereignty and that they must be treated as a "common concern of all mankind." The "*inter*" and "*intra*" as part of the reigning principles deserves to be emphasised as they underscore the onus to ensure equality in the use of natural resources, not just between and among members of different generations, but among members of *the same generation*.

3.0 The Tragedy and Status of Declining Forests

On a global scale, forests are among the most "imperilled" or endangered natural resources. It has been observed that most of the forest decline takes place in developing countries, particularly in tropical areas, and for the avoidance of doubt:

> *the process generates large amounts of carbon dioxide, equivalent to 20% of global emissions from fossil fuels, making deforestation the second most important contributor to global warming and results in annual degradation of some 12 million hectares of fertile land and loss of thousands of species (estimates range between 8,000 and 28,000 per year). Deforestation and forest degradation directly threaten as many as 400 million people, including 50 million forest indigenous people who depend on forests for subsistence.*
> (Contreras-Hermosilla, Arnoldo 2000: 5)

Other estimates have stated that up to 50% of tropical wood imported into Europe comes from "illegal" sources resulting in losses of up to *US$15 billion* per year in tax revenue for the would-be beneficiary countries.[6] It has been reported that a three-week forest fire in Indonesia in 2015 was the source of more Green House Gas (GHG) emissions than the whole of Germany in a year, amounting to some *US$ 166 billion* worth of losses to forestry, agriculture tourism and other industries; not to mention the human death toll of no less than 100,000 people.[7] Unlike many other contemporary environmental concerns, the global community is yet to come up with a more stringent and legally binding agreement that can effectively promote protection and conservation. Part of the reason is that global agreements are canvassed and agreed in environments and in terms of member-state representation (the "CoPs") that clearly fail what one might call, "the tracking test."[8] It, therefore, remains a major moot point if these solemn conventions; often prefaced by some very high-sounding "*chapeaus*" can actually provide impetus and direction for handling the global complexity around environmental degradation generally and forestry decline in particular. At issue in this context is the aforementioned NLBA, 2007 that in many ways, typifies the ineffectiveness being referenced here.[9] Part of the "creative legal solutions," one believes, should be teased from the already emerged corpus of International Environmental Law (IEL). They must, thus include, some of the long-standing international law norms; the legal decisions and advisory opinions of the International Court of Justice (ICJ) as well as, the considerable body of international conventions and treaties that collectively address the world's environment-related challenges, some of which we have cited in a separate section. Instruments such as the CBD and CITES, and even more remarkably, the *Convention on the Law of non-navigational use of International Water courses* (1997)[10] have already enumerated certain broad principles that one believes, have a good bearing on forest protection. For the purposes here, there are just two approaches that could represent the promising and effective solutions to the forest decline quagmire.

4.0 Two Existing Forestry Protection Approaches That Are "Creatively" Effective

Given what we have stated in the foregoing sections, it is necessary to preface this section by acknowledging that the effectiveness of most – if not all – international environment agreements are under some form of siege. To declare that there are, in fact, some existing instruments that are entirely effective and delivering as expected, would probably be an overstatement. Some previous analyses have clearly indicated that it is usually rather difficult to assess "effectiveness," especially, when much of the reliance data and information is often generated by the same governments and other "implicated institutions," which then casts aspersions on their integrity.[11] This notwithstanding, we suggest that there are some "creative" and worthwhile efforts, particularly those emanating from the non-multi-lateral processes. In this regard, in terms of our notion of "creative solutions" there are at least two promising approaches to date that are worth commenting upon, namely:

(i) The *Climate, Community and Biodiversity Alliance* (CCBA) standards,[12] and;
(ii) The aspirations contained in the *International Tropical Timber Agreement* (ITTA 2006).[13]

Whereas the CCBA is clearly an effort largely driven by non-Governmental interests, the latter, ITTA stands out as an effort of governments through multi-lateral processes spearheaded by the UN, and which possibly offered more opportunity for "effectiveness" than what the international community came up with later. The CCBA represents, "the most widely used and, indeed, the most respected international standards."[14] The main reasons why the CCBA standards are considered most effective are partly self-evident from the noble aims declared by the promoters.[15] One is that they target "high-quality and *multi-benefit*"; (i.e., not a single benefit) land-based carbon projects. They also must go through some rigorous procedures and processes; notably, a validation by external auditors and evaluators as well as a mandatory public comments stage. It goes without saying that these basic good-governance imperatives have been the biggest shortcoming of some agreements. Moreover, the CCBA standards have put together implementation toolkits that guide implementers and they have specified some very specific deliverables that are verifiable; monitorable and assessed on a periodical basis. Furthermore, the CCBA has set its own targets regarding GHGs reduction (4.4 million tons of CO_2 as of 2008) and a forest cover target through re-forestation of *1,385,190 hectares*. It is apparent that it is for these rather stringent standards that many developed country-based investors who want to "ethically engage," have found the CCBA approved projects attractive hence higher uptake than those under the Kyoto Protocol's CDM mechanism.[16]

On the other hand, the ITTA claims a membership that represents the bulk of the world's tropical forests and all the leading players in the tropical timber trade. It has an implementing body [the International Timber Organization (ITTO)] charged with the responsibility to oversee the implementation of the agreement. The ITTO states in its 2013–2018 strategic plan that its aim is to promote the conservation and sustainable management, use and trade of tropical forest resources. It develops internationally agreed policy documents to promote sustainable forest management and forest conservation and assists tropical member countries to adapt such policies to local circumstances and to implement them in the field through projects. The ITTO collects, analyses and disseminates data on the production and trade of tropical timber and funds projects and other actions aimed at developing industries at both community and industrial scales and has, in this respect, declared that it has so far funded over 1,000 projects valued at around US\$ 350 million since it became operational in 1986.[17] One other positive about the ITTA is that it has been responsive to periodic revisions and amendments since it first became operational in 1983 and thus ensuring some kind of response to changing global realities.

4.1 Four Other "Creative" Solutions and Approaches

The notable absence of a stronger enforcement and legally binding regime of forestry protection has not been the consequence of lack of effort or the dearth of ideas. Rather, as already hinted, there exists quite a huge body of measures and approaches; including to date, more than *1,300* Multi-Lateral Environment Agreements, (MEAs); over 2,200 Bilateral Environment Agreements (BEAs) and over 90,000 individual countries "membership actions," according to this very source.[18] Accordingly, it is the international enforcement of "the better ideas" in order to secure a more certain lifeline for forests that is perhaps the essential challenge. This challenge at enforcement is fuelled by two main factors. One is the international convention around *state sovereignty*, which by practice, places the greater compliance responsibility on state parties. Second, is the practice of placing "developed countries" on one side, and "developing countries" on the other. We argue here that as a bare minimum, it is this "single binary" or "Manichean categorization,"[19] with its roots in the contested history of both slavery and colonialism that has to be confronted. At the same time, there has been well-noted difficulty in transferring capital, technology, and capacities in order to enable those designated as "developing countries," to fully comply with their obligations under the declared principle of "*common but differentiated responsibilities.*"[20] In seeking and emphasising creative approaches, one is not necessarily talking about entirely new inventions: rather, within the corpus of existing instruments, there is room for creativity and enforcement in ways that will certainly improve the forestry governance regime. The four other creative approaches discussed subsequently, provide pointers which can be immensely beneficial *if applied correctly* and *stringently*. The four are:

1) Delimiting *state sovereignty* and expanding *international jurisdiction* over forests; including the introduction of the offence of "*ecocide.*"[21]
2) Expanding the scope of the already well-established principle of, "*Obligation not to Cause Significant Environmental Harm.*"
3) Re-conceptualising the "*Common but Differentiated Responsibilities*" doctrine, and;
4) Strengthening the role and participation of *Non-state actors* in forest protection.

4.1.1 Delimiting State Sovereignty and Expanding International Jurisdiction

There can hardly be any disagreement that just like waterways, oceans and the ecosystems in general, forests affect climates further afield than just the immediate political boundaries. To put it more succinctly, forests qualify as a transboundary resource. As was eloquently expressed way back by the World Commission on Environment and Development:

> *the traditional forms of national sovereignty are increasingly challenged by the realities of ecological and economic interdependence. Nowhere is this more true than in shared ecosystems and in the global commons.*
> (1987: 261)[22]

Part of the solution that one is recommending is therefore a derivative from similarly situated transboundary resources. For instance, **articles 5(1); 7 and 20** (to name just a few relevant clauses) of the *Convention on the Law of the Non-navigational Uses of International Watercourses* (1997) are emphatic on the imperative of, "taking into account the interests of the *watercourse states* concerned, consistent with adequate protection of the watercourse" (5(1).[23] The more stringent application of both a wider (global) and a domestic jurisdictional principle makes sense because the "jurisdictional hook" exists already under "subsidiarity" and "complementarity" garnered from international criminal tribunals lately domiciled under the International Criminal Court's Rome (1998) statute. The personal jurisdiction premise would strategically aim at ensuring that liability for forest-related crimes shifts from the rather abstract notion of "state responsibility" to a more individualised/personal accountability, including the possible introduction of the offence of ecocide.[24] Under the offence of ecocide, we would then begin to see real possibilities of dealing with the crooked, capitalist elite that continue to plunder rain catchment areas and major water towers, such as the Mau Forest in Kenya; the majestic Congo basin forest, and of course, the Amazon.

In turn, the universal jurisdiction would be consolidating the reality that forests are indeed part of the "global commons" and there is justifiable basis for a collective exercise of jurisdiction in a fully criminal sense. If *ecocide* is successfully introduced by a member state as part of the International Criminal Court's jurisdiction, it simply means that personal liability for crimes committed against forests and the environment is generally extended. The international regime would then specify more stringent prohibitions against abuse of forest resources; must have prosecutorial powers and be able to initiate some level of international monitoring of the timber trade, such as application of "jurisdictional zones"; application of similar principles to shared watercourses, i.e., "Prior Notification;" "Fair and Equitable Use;" and "Prevention of transboundary Harm"). To put it differently, just like certain categories of crimes that have been adjudged as belonging to be "against humanity," there is a strong argument to be made for international "crimes against the environment."

4.1.2 Expanding the Scope of "Obligation Not to Cause Significant Environmental Harm"

There is probably no other IEL principle, approach, or standard that nears the universal consensus triggered by this particular obligation. From Stockholm (1972) to Rio (1992) and the Paris Agreement (2015), the most influential moral obligation has been the prohibition targeting the use of the environment by anyone in ways that cause significant harm to others. The most famous ICJ decisions have revolved around the use of transboundary water resources in ways that do not harm the interests of downstream states.[25] The distinct positive about this obligation in so far as forests are concerned lies in its straightforward clarity and simplicity, which means that there can hardly be any claims of ambiguity or uncertainty. What remains must therefore be the creative ways and means of ensuring that this principle survives in theory and in practice as a cardinal responsibility in the call for the protection and conservation of forests. As is quite evident, in many of the UN-mediated agreements, including the "Paris agreement," the crimes or the harms have not been clearly defined; neither has been, the levels of liability or damages applicable. The NLBA, for instance, is spectacularly distinct for its listing as a voluntary instrument, implying that it leaves all undertakings to the supposed goodwill of the sovereign state. Whereas this may have given then-President Obama the leeway to assuage pressing domestic concerns within the USA,[26] the argument here would be that "*voluntary*" and "*non-legally*" binding agreements cannot work effectively, especially where there are very weak governance and Rule of Law imperatives. We have thus attempted, for illustrative purposes, to re-draft, or better still, introduce some four key articles that could strengthen the NLBA and hopefully, provide for more stringent state party obligations.[27]

4.1.2.2 Redrafting Article 2 of the Forest Instrument

PROPOSED HEADING: SOVEREIGNTY OVER FOREST RESOURCES

Instead of Article 2, which originally headlines "Principles," we propose "Sovereignty Over Forest Resources" as headline. Under sub-article (b) which reads, "*Each State is responsible for the sustainable management of its forests and for the enforcement of its forest-related laws,*" we would propose an amendment:

> *States have, in accordance with the principles of international law, the sovereign right to [responsibly] exploit {their own forest} resources pursuant to their own environmental and developmental policies, and the responsibility to ensure that activities within their jurisdiction and control do not cause damage to the environment of other states or of areas beyond the limits of national jurisdiction.*

4.12.3 Redrafting Article 5

PROPOSED HEADING: SUSTAINABLE FOREST MANAGEMENT AND BENEFIT SHARING

In place of the *Article 5* headline reading, "National policies and measures" with a whopping 25 rather rumbling sub-articles, we would propose the more all-embracing, "*Sustainable Forest Management and Benefit Sharing*" and having a more transboundary-mindful, simplified clause:

> *The developed country Parties shall endeavour to assist the developing countries in compliance with their obligations to sustainably*

manage forests through the provision of funds, technical assistance, capacity building and relaxed compliance schedules. Developed and developing country Parties will strive to monitor operations of forest-related companies operating in nations other than the nations where they are headquartered. Parties to the Convention will endeavour to work towards equitable distribution of the benefits from forest products.

4.1.2.4 Introducing an Article 8: Indigenous Peoples, Forest Dwellers, and Local Communities

In order to strengthen the contribution of indigenous communities and allay the fears of local communities and forest dwellers, given the immense capital power of the global "tenderpreneurs" in forestry, we propose an *Article 8* headlined as above and providing that:

States Parties recognize the necessity and long-term benefits of full involvement of indigenous peoples, forest dwelling peoples, and local communities in the sustainable management and conservation of forests. Accordingly, States Parties shall pursue appropriate means to ensure the participation of all indigenous peoples, forest dwelling peoples, and local communities in all forest-related programs, projects, and activities at the national and local levels. All parties agree to assess the environmental and social impacts of any forest-related projects that threaten the interests of indigenous peoples, forest dwelling peoples, and local communities. All such environmental and social impact assessments shall be developed through processes that allow the participation of all affected indigenous and forest dwelling communities, and all community stakeholders.

4.1.2.5 Introducing an Article 10: Endangered Species

A key consideration in sustainable forest management concerns the fate of endangered forest species. Although this aspect may have been taken care of under CITES, its reiteration in the forestry instrument is considered vital. We thus propose an Article 10 providing special attention to endangered forest species:

State parties shall commit to taking every possible care and precaution to ensure the safety and livelihoods of all endangered species of trees or of any forest dwelling animals. To this end, state parties shall designate and preserve clearly specified forest areas that are known habitats for such species in the form and quality necessary to sustain the species for both present and future generations.

As can be seen in the attempt to re-draft the three articles above, a more effective approach must get state parties to agree to more strengthened language and to give more legal force to the forest instrument. Most of all, there is much to learn from Chester Bowles's assertion that speaks for itself:

> *20 percent of countries would normally comply automatically; 5 percent would evade compliance but . . . 75 percent will comply if the 5 percent can be caught and punished.*[28]

4.2 Re-conceptualising "Common but Differentiated Responsibilities"

The prevailing mantra has clearly over-privileged a recurrent "suspect" classification. At one level, *developed vs. developing* countries tends to engender perpetually unrealistic expectations, including fomenting stereotypes. In the absence of institutions, and expertise for reliable data mining and processing, one has to question how DR Congo would ever justify a *US$ 30 billion* request submitted to the Green Climate Fund.[29] On the other hand, the annual *$100 billion* pledge under the Paris Agreement has been met with practical difficulties relating to actual access to the funds but also no guarantees for prudence and responsible use. Further afield, one report indicated that Indonesia (probably the world's largest source of tropical timber) received only *$60 million* in 2015 against a supposed pledge of US$ 1 billion "if Indonesia stopped cutting its trees."[30] More revealing is that a presidential moratorium including a poorly-evidenced commitment to reduce GHG emissions by 29% faltered. At a glance, decisive presidential actions, usually called "political commitment" in many places, are probably welcome but obviously not always sustainable. In the meantime, with Brexit and the unprecedented Covid-19 pandemic which literally paralysed the entire world, one has to worry if the huge financial commitments from developed countries will always prevail.

In re-conceptualising what might well be [re]-phrased as *the common responsibility for all mankind*, again it is only tempting to count on President Biden's election in 2020 on the democratic ticket, emphatically suggesting a major shift in the American policy and attitude towards environmental conservation and protection. America's leadership would certainly not be averse to a re-conceptualisation that sees all state parties taking their respective roles much more seriously on the basis of independent and "responsible sovereignty" as aptly opined by one former Secretary of State.[31] The bottom line, according to this line of thinking would be that sustainable forest management cannot be a mere appendage of the wider national development effort. Rather, it implies more targeted support for building blocks such as democratic reforms and more broad-based institutional capacities in which Information, Education, and Communication (IEC) for greater public awareness is cardinal.

4.3 Strengthening the Role of Non-state Actors

On paper, some current IEL approaches and standards that have emphasised the principle of "*Joint Management*" and other multi-stakeholder involvement approaches in forest protection programmes are definitely worthwhile. Getting local forest communities to be somehow involved in decisions that relate to forests is important, just as much as the engagement of business and private sector communities that are part of the international trade in forest products. We picked on the ITTA and the ITTO in the foregoing sections because these are precisely what these initiatives have attempted to accomplish. The CCBA standards are especially unique as they put a premium on all these levels of engagement. However, the downside is that an effective public participation process or "joint management" is still very much dependent on government-led processes. Ordinary citizens in nearly all cases, tend to be mere consumers of what governments think. Moreover, private sector voices would normally be much more powerful than those of poor forest communities with the result that, in spite of the best of intentions, the wishes and intentions of these dominant voices would be the ones to carry the day. For progress to be made, there has to be genuineness in really trying to tap into what poor, indigenous communities know about forests and their possible husbandry.

5.0 Lessons From Creative Literatures: "Things Fall Apart"

Chinua Achebe's pioneering novel – *Things fall Apart* – remains one of the most discussed pieces of creative literatures to have emanated from the Continent of Africa. The numerous critical perspectives have not only canvassed the incredible fatality associated with the author's characterisation of Okonkwo as the main protagonist, but they have also numerously un-packed meanings that one would make of the traditional culture and lifestyles of the Ibo of Nigeria; the impact of Western colonialism on those highly held traditional African values and beliefs; the political implications of the novel, and in this instance, (along with a few pace-setting others), the dimension of "environment" not just in *Things Fall Apart*, but also in his other well-known works, notably; *Arrow of God*.[32]

The purpose here, by way of concluding our analysis, is a little more modest and only serves to illustrate what, in celebrating the life of Achebe, Elaine Savory correctly characterised as, Achebe's "eco-critical awareness." It speaks to our broad thematic focus around the "Environmental Humanities," and as Savory asserts:

> *Hardly noticed, however, has been his insightful and important representation of Igbo culture and modern Nigeria in ecological terms. As we mourn him, those of us who think about ecology are anxious that human beings might one day have to mourn the loss of prime habitat, perhaps even the whole earth as habitable space. Achebe profoundly*

> *understood the consequences of losing a sense of balance between people and their environment long before environmental concerns became a movement and a consciousness.*
>
> (Savory, Elaine, (2014), with our emphasis)

It is to these largely unexplored aspects of TFA that we must now turn, however briefly, in order to demonstrate some powerful, if not sacred ancient lessons around forests in particular and environmental justice generally.

5.1 The Sacred Sanctity of Forests in African Cosmology

Those familiar with TFA as a pioneering African literary gem will already know that its primary power lies in the manner and style in which it portrays the near-infallible order of the pre-colonial African society. By the end of the novel that idyll society is completely disrupted by the advent of Western colonialism. TFA is substantially plotted around the tragedy of Okonkwo, a traditional Ibo man whose own lifetime changes and the personal tragedy he undergoes, symbolises the progressive historical disruptions that many an African society has had to put up with since the advent of colonialism. In the beginning, we are treated to one of those significant incidences that underlines the Ibo man's harmony and interlinkage with his environment and to nature. Okonkwo's father (Unoka) and his legendary laziness is what results in the inability to harvest anything, and hence his son's zero inheritance and the consequential cyclic of poverty and want which Okonkwo does everything in his power to escape from. Unoka is well-known in the entire clan for tilling exhausted lands instead of marshalling his efforts into what is aptly described as "virgin forests," and hence the early encounter between him and the Priestess Agbala speaks for itself:

> *You have neither offended the Gods nor your fathers. And when man is at peace with his god and his ancestors, his harvest will be good according to the strength of his arm. You Unoka are known in all the clan for the weakness of your machete and hoe. When your neighbours go out with their machetes and axes to cut virgin forests, you saw your arms in exhausted farms that take no labour to clear. They cross seven rivers to make their farms; you stay at home and offer sacrifices to a reluctant soil.*
> (TFA 1958/1996: 13)

One key lesson here is the importance of ploughing "virgin forests" which, in turn, leads to assured production and good harvests. As crucial is the allusion to "sustainable forest management" which does not simply imply that forests cannot be used to satisfy human needs; in fact, on the contrary, they have to be somehow harnessed and exploited. In the international forest protection regimes already discussed, this would be akin to what many countries see as "the sovereign right to development." Yet, Achebe is simultaneously calling attention to the fact that when farms get exhausted as a result of over-use, the time-tested indigenous wisdom is that there has to be a period of rest to enable the soils to replenish. The creative legal solution that is relevant here is the idea of "forest quotas," and protected conservancies earmarked for deforestation and pledged reforestation, implying that when one part of the forest "rests" the other can be harnessed and exploited.

Later in the novel, Achebe presents what must be the most significant environmental catastrophe with lessons relevant today:

> *the year that Okonkwo took eight hundred yams from Nwakibie was the worst year in living memory. Nothing happened in its proper time; it was either too early or too late. It seemed as if the world had gone mad, The first rains were late and when they came lasted only a brief moment. The blazing sun returned more fierce than it had ever been and scorched all the green that had appeared with the rains. The earth burned like hot coals and roasted all the yams that had been sown.*
> (TFA 1958/1996: 16–17)

Subsequently, when the rain came, "*it rained like it had never rained before. trees were uprooted and deep gorges appeared everywhere.*" (TFA, 17). Two points are worthy of note from this passage. The first is that the unpredictable climate that destroys everything in its wake is clearly signalling to what today passes as global warming and climate change that has been going on for quite a while. Second is the principle of "sharecropping" that enables Okonkwo to rebuild his own barn of yams. "Sharecropping" implies that after the farmer has toiled, he is only entitled to one-third of his efforts because of the inherent obligation to pay whoever lent the seedlings. At the same time, the sense of community is paramount because even amid this tragedy Okonkwo still has onerous responsibilities including taking care of his own father's household.

In the entirety of TFA, there are numerous allusions to the sanctity of forests. It is to be remembered that whenever the mortals go into consultation with the oracles, the priestesses such as Chika and Chielo enter into that unfathomable spiritual domain. The shrines are inside the forest and can seemingly only be accessed through a tiny hole at the side of a hill (TFA 1958/1996: 12), in which case they become almost unrecognisable. The village playground (the ilo) is an open ground in the midst of the forest sanctified as a place where the community lives; celebrates, makes important decisions, and confirms the important ideal of community. Not lost on us is the fact that practically every Umuofia man carries along his stool, presumably a forest product whenever there is a public function. An interesting role of the forest is that it is at times represented as the embodiment of evil; where the untoward, such as the undignified death of Unoka must find repose and similarly so, the tragic end of Okonkwo. Yams, so we are told, are a major signifier of the Ibo sense of wealth; as it were, not only "the king of crops" but also the symbol of "manliness, and he who could feed his family on yams from one harvest to another was a very great man indeed" (TFA, 1958/1996: 23).

Without belabouring the point, it is clear that in this setup, the forest ecosystem is graduated, and everyone knows the ranking of every tree or crop. The same concept can be extended to the kola nut as well as the palm tree, which by tradition, have such strict, if not sanctified protocols. Thus, a woman, cannot on her own motion harvest the palm tree even though it takes the symbolic violation by Ekwefi (Okonkwo's younger wife) to introduce a thread of an emerging revolution by the womenfolk. The silk-cotton tree is where "spirits of good children lived, waiting to be born" (TFA, 33). And when men like Okonkwo take their spiritual and godly role of supreme court justices (the "egwugu"), they actually live in forest-like shrines into which no other mortals dare to go. Not lost on readers of TFA is the series of taboos and sanctions used and deployed in preserving the forest, leaving no doubt that they designated the space between the living and the dead. There was coming and going between them, and most graphically confirming life itself as a continuum of going and becoming, hence a harmony in that lost past, between humans and the total ecology of existence. Last but not least, forests are also a major source of medication and that's where the parents run to first when the young Enzima falls sick.

Even without taking it much further, a nuanced analysis of TFA confirms one or two principles of relevance to the creative enterprise of securing the environment and forestry at large. Implicit here is the sense of community that now finds expression in such doctrines as "Joint Management" as well as the special and necessary jurisdictional zones that enable a staggering of plantations. Moreover, it is clearly critical that forest ecosystems must be ranked and have their various intractable values appreciated.

Notes

1 An initial draft of this paper was presented during the 'Environmental Humanities Colloquium,' held at the University of Pretoria, South Africa in May 2019. It has since been revised for this publication.
2 This is one of the oft-quoted cardinal principles providing the non-contestable basis for practically all of the adopted international environment protection instruments.
3 Since adoption in 1992, The UNFCCC has had many other 'protocols' such as the 'Kyoto Protocol (1997) and other subsequent agreements in the global effort to enhance adaptation and mitigation of climate change.
4 See, https://undocs.org/pdf?symbol=en/A/RES/62/98 also popularly referenced as 'the Forest Instrument'.
5 For an idea of the current status, see University of Oregon data base: https://iea.uoregon.edu/iea-project-contents.
6 These are estimates in Hunter et al. (2011: 1166) but the figures could well be much higher today.
7 For this, "the Burning question" a special report on Indonesia in *The Economist*. November 26, 2016.

8 The main point is that in these 'Conference of the Parties' (CoPs), it is the Member State, however abstract or weak, unrepresentative and often lacking professional knowledge and experience, is the recognized legal actor.
9 Section 4.1.2.2 provides some ideas on how the UN 'Forest instrument' can be strengthened.
10 See https://legal.un.org/ilc/texts/instruments/english/conventions/8_3_1997.pdf
11 As examples only: "Assessing the Effectiveness of International Environmental Agreements: the Case of the 1985 Helsinki Protocol" by Evan J. Ringquist and Tatiana Kostadinova in, *the American Journal of Political Science*. Vol. 49, No. 1 (Jan., 2005), pp. 86–102; and Kellenberg and Levinson Arik, "Waste of Effort? International Environmental Agreements" in, *Journal of the Association of Environmental and Resource Economists*, March/June 2014).
12 For the specific CCBA standards, see: https://verra.org/project/ccb-program/
13 For the Full text of the ITTA, see: https://treaties.un.org/doc/Treaties/2006/02/20060215%2004-26%20PM/Ch_XIX_46p.pdf.
14 With emphasis from Prof Takacs: IEL discussions, November 24, 2016.
15 CCBA is an alliance that seems to embrace the participation of the bigger non-governmental interests and players rather than governments *perse*; notably, CARE, Rainforest Alliance, The Nature Conservancy, Wildlife Conservation Society, BP, GFA Envest, Intel, SC Johnson, Sustainable Forestry Management Ltd., Weyerhaeuser, and advising institutions.
16 The Kyoto Protocol (1997) CDM under its article 12 provides 'credits' for projects, but evidently the uptake was much lower.
17 See, the ITTO Strategic Action Plan, 2013–2018. www.itto.int/files/user/pdf/publications/ENGLISH_ACTION_PLAN_2013_2018.pdf.
18 For an idea of the current status, see University of Oregon data base: https://iea.uoregon.edu/iea-project-contents.
19 This is generally part of the anti-colonial narrative earlier conceptualized in these two opposing terms and indebted more famously to the writings of such personages as Frantz Fanon (1963) *the Wretched of the Earth*.
20 Just like the earlier referenced principle of 'inter and intra-generational equity,' this too is a well-established principle under the UNFCCC (1992) and the Paris Agreement (2015), all emphasising the onus placed on developed countries to avail more resources for implementing climate and environmental protection in the less-endowed 'non-Annex 1/developing countries.
21 For the full elucidation of the notion of 'ecocide I am indebted to the illuminating advocacy of Femke Wijdekop during the 2017 *International Environment Laureate Convention*. Freiburg, Germany, (9th to 12 March 2017).
22 See, *Our Common Future* (1991).
23 For references to this particular agreement, please see: https://legal.un.org/ilc/texts/instruments/english/conventions/8_3_1997.pdf.
24 For an elucidation of the remits of 'ecocide', I am grateful to Ms Femke Wijdekop for insights on the concept at the International Environment Laureates Convention (Freiburg, Germany, 2017).
25 I am referring here to the 1997 ICJ judgment: Gabcikovo-Nagymaros (*Hungary v. Slovakia*) and the 2010 *Argentina v. Uruguay Pulp Mills*. As relevant is the 'Trail Smelter' (1941) where Canadian fumes caused damage to US citizens and, the 'Corfu Channel' (British war ships damaged in Albania waters).
26 The main 'pressing domestic concerns' every American president must contend with in the foreseeable future is always the demands of powerful business and industry lobbies who, as it is probably obvious, are most culpable when it comes to GHG emissions and general environmental pollution.
27 For the wording and suggestions contained in the four articles referred here, I duly acknowledge the immensely interesting discussions and insights provided

by the entire IEL class of 2016 at the UC-Hastings Law School, USA under Prof David Takacs.
28 Chester Bowles as quoted in Hunter (2011: 399), with our emphasis, while for the redrafting of articles, I am grateful to all UC-Hastings (LLM 2016 class) for their in-puts and deliberations.
29 This figure is reported in other sources and the main UNFCCC website: http://www4.unfccc.int/submissions/indc/Submission%20Pages/submissions.aspx
30 See, *The Economist*, November 26, 2016: 34–35.
31 'Responsible sovereignty' can be attributed here to the Hon Dr Condoleezza Rice on the occasion of the W.E.B Dubois Lecture Series at Harvard University, Jan 10, 2017. See www.youtube.com/watch?v=DLdgi1m6lkg.
32 For an illuminating additional 'eco-critic' of Achebe's (1964) TFA, see Barau, S (2009).

References

Achebe, C. 1958/1966. *Things Fall Apart*. London: William Heinemann Ltd Publishers. *https://www.amazon.com/Things-Fall-Apart-Chinua-Achebe/dp/0385474547*
Achebe, C. 1964. *Arrow of God*. London: Heinemann.
Arnoldo, C. 2000. *The underlying causes of forest decline*. Occasional Paper No. 30. Jakarta: Centre for International Forestry Research (CIFOR).
Barau, S. A. 2009. Bridge-building between literature and environmental values of Africa: Lessons from things fall apart climate, community and biodiversity alliance. *https://verra.org/project/ccb-program/*
Convention on Biological Diversity (CBD). 1992. *www.cbd.int/doc/legal/cbd-en.pdf*
Convention on International Trade on Endangered Species (CITES). 1973. *www.cites.org/sites/default/files/eng/disc/CITES-Convention-EN.pdf*
Convention on the Law of Non-navigational Use of International Water Courses. 1997. *https://legal.un.org/ilc/texts/instruments/english/conventions/8_3_1997.pdf*
Fanon, F. 1963. *The Wretched of the Earth*. Harmondsworth: Penguins.
Hunter, D. et al. 2011. *International Environmental Law and Policy*. New York: Foundation Press/Thomson Reuters.
International Tropical Timber Agreement. *https://treaties.un.org/doc/Treaties/2006/02/20060215%2004-26%20PM/Ch_XIX_46p.pdf*
Kyoto Protocol. 1997. *https://unfccc.int/resource/docs/convkp/kpeng.pdf*
Kellenberg, D. and Levinson, A. 2014. Waste of effort? International environmental agreements. *Journal of the Association of Environmental and Resource Economists* (JAERE), 1:135–169.
Ringquist, E. J. and T. Kostadinova. 2005. Assessing the effectiveness of international environmental agreements: The case of the 1985 Helsinki protocol. *The American Journal of Political Science* 49:86–102.
Savory, E. 2014. Chinua Achebe's ecocritical awareness. *PMLA* 129:253–256.
Stockholm Declaration. 1972. *http://docenti.unimc.it/elisa.scotti/teaching/2016/16155/files/file.2017-03-11.7227158899*
Takacs, D. 2016. *International Environment Course Outline for the LLM Programme*. San Francisco: UC Hastings, College of the Law.
The Economist. 26 November 2016.
The Paris Agreement. 2015. *https://unfccc.int/files/essential_background/convention/application/pdf/english_paris_agreement.pdf*

The Rio Declaration. 1992. *www.jus.uio.no/lm/environmental.development.rio.declaration.1992/27.html*
The UN Framework Convention on Climate Change. 1992. *https://unfccc.int/resource/docs/convkp/conveng.pdf*
The UNs Non-Legally Binding Instrument on All Types of Forests. 2007. *https://undocs.org/pdf?symbol=en/A/RES/62/98*
The World Commission on Environment and Development. 1987. *Our Common Future*. Oxford: Oxford University Press.
Vienna Convention for the Protection of the Ozone Layer. Vienna. 1985. *https://treaties.un.org/doc/Treaties/1988/09/19880922%2003-14%20AM/Ch_XXVII_02p.pdf*

11 Carbon Dioxide, Climate Change, and an Energy Transition for a Future Africa[1]

Emil Roduner and Egmont R. Rohwer

1. Introduction

Burning fossil fuels since the beginning of industrialisation about 250 years ago has led to an increase in atmospheric CO_2 concentration with concomitant temperature rise considerably beyond any levels over the past 800 000 years, as determined from Antarctic ice core analysis (Lüthi et al. 2008; Brook 2008). Considering the fact that people have adapted to live at temperatures between +50°C and −50°C, an increase of the average global temperature by 2°C as envisaged by the Paris Agreement on Climate Change (Paris agreement 2015) looks negligible. By comparison, the preindustrial temperature fluctuations over a period that covered several glacial areas amounted to ca. 8°C. An increase of 2°C is therefore expected to have severe consequences for life on Earth, with a sea-level rise by several meters and changes in habitability and biodiversity due to climatic changes, expanding deserts, and disappearing glaciers.

There are worldwide efforts towards a net-zero carbon emission economy by replacing fossil fuel with renewable energy, mainly solar photovoltaics, and wind electricity. This implies a worldwide, fundamental energy transition (Editorial: Coal in a hole 2019). Beyond this, CO_2 can be captured and stored, or utilised as feedstock for renewable liquid solar fuel that can be traded or used for energy storage.

More than 80% of the African population is not connected to an electricity grid. Solar and wind energy allow for a significant improvement of living standards in off-grid locations. Electrical power is needed for water purification, desalination, and recycling, and for the operation of refrigerators to preserve medical drugs and keep food fresh. It permits connecting to the world while preserving tribal entity, local culture, language, and religion. This will reduce significantly any migratory pressures.

2. North-eastern South Africa: A Mining Paradise

The area of the ancient inland lake in the Witwatersrand basin, around Johannesburg, the city that originates in the 1886 gold rush, is famous for its tremendous deposits of valuable minerals (Figure 11.1). The gold

DOI: 10.4324/9781003287933-11

Carbon Dioxide, Climate Change, and an Energy Transition 171

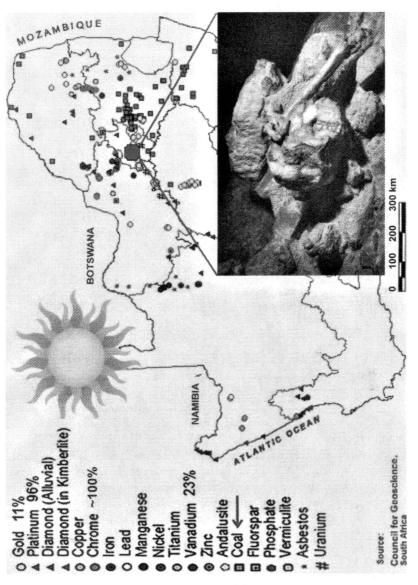

Figure 11.1 South African mining map with an indication for the Cradle of Humankind (World Heritage Site, large grey symbol indicated by two connectors).

Source: (colour online, amended from Utembe et al. 2015)

reserves are estimated to amount to 11% of the world's gold. Moreover, the nearby Merensky reef, a solidified magmatic layer of only 1–3 m thickness, contains 96% of the world's platinum that has a price on the order of 2/3 that of gold. Platinum is a precious catalyst and occurs in the reef together with the other precious platinum-group metals (palladium, rhodium, iridium, and osmium, which are all more expensive than gold). In certain years, the mined chromium reached nearly 100% of the amount mined worldwide. Vanadium resources are rated at about 23% of the world's resources, and there are large amounts of iron and manganese as well as other metals. In addition, there are major occurrences of diamonds, and there is coal for another 200–300 years at the present rate of mining.

Sunshine is not a mineral, but a treasure with abundance, on par with that experienced in Australia and at least a factor of two more than for example in Germany, a pioneering country regarding the use of solar energy.

Right in the centre of this paradise of treasures is the Cradle of Humankind, declared a World Heritage site, where a number of 2–3 million years old hominin fossils, notably of *Australopithecus africanus*, were discovered. It reminds us of the biblical story of creation, in which God places Adam and Eve in the Garden of Eden, in the middle of all the treasures that he created. He blessed them and said to them,

> *Be fruitful and multiply and fill the earth and subdue it and have domination over the fish and over the birds and the heavens and over every living thing that moves on the earth.*
> (Genesis 1:28)

This text may be seen as the origin of an anthropocentric view in particular of the western world of Christian tradition that places man above everything else, prompting him to utilise available resources. It mentions only the living creatures and not the minerals, but its uncontrolled interpretation may nevertheless have contributed to the irresponsible use of this wealth with all its environmental and social consequences.

The anthropocentric worldview stands in pronounced contrast to the world view of many indigenous religions in Africa and in North America in which humans live in harmony with Mother Earth, with animals and plants, preventing its excessive exploitation.

3. Carbon Dioxide Emission and the South African Energy Transition

Africa has made a minor contribution to the accumulated fossil carbon dioxide emission since 1850 and is overall still not a very significant contributor (Winkler 2018). South Africa, however, has had a strong increase over the past decades, resulting in an emission of 468 Mt per annum (1917), which corresponds to 1.3% of the world's emission and places the country on rank 14 among heavy emitters. About 65% of this emission has its origin in point sources, with 50% from Eskom's coal-fired power plants and Sasol's fuel production from coal counting as the world's largest point emission source.

The most straightforward and cheapest way to reduce South Africa's carbon footprint comprises a fundamental energy transition from coal to renewables, mostly solar photovoltaic and wind energy. So far, the country has made little use of its abundant solar irradiance (Solar GIS 2011), but the new version of the Integrated Resource Plan outlines a significant policy change and points out that solar and wind energy are cheaper than electricity from coal or nuclear origin (IRP 2019). Moreover, a study of aggregated wind and solar photovoltaic potential revealed that 50% of the nation's electricity demand can be produced from these renewable sources without any additional storage capacity. Compared to, for example, Germany, the complementarity of the two sources were found to be far better in SA where the installations are dispersed over a large area and take advantage of the available country-wide electricity grid (Knorr et al. 2016). In Germany, the fraction of electricity that is being produced from solar and wind energy has exceeded 30%. On sunny days, the renewable electricity produced exceeds the total demand. The excess power is then used for the electrolytic production of hydrogen that is simply added to the large available gas grid. Its combustion produces water, and the mixture thus reduces the amount of carbon dioxide emission when compared to heating with natural gas alone.

Significant opposition against the proposed energy transition comes from the union of coal miners who fear the loss of their jobs. This fear is understandable but has to be seen in the context of the new jobs arising for engineers in the renewable energy sector. The massively increasing number of university graduates from various backgrounds are unlikely aiming to work as miners. Moreover, the transition will not happen overnight but will take several decades. An issue is the clustering of the lost coal mining jobs in a relatively small region (Figure 11.1), which requires being addressed early by the government to avoid high local unemployment.

Although some of the oldest coal-fired power plants are planned to be phased out soon, the two large new installations at Kusile and Medupi, which are only just opening up, are expected to produce electricity and emit carbon dioxide for several decades to come. A solution is needed to cope with this situation. Current intense and worldwide efforts aim to capture carbon dioxide at such point sources or even from the atmosphere and recycle it, using renewable energy, to liquid fuels (Käthelhön et al. 2015). Such fuel counts as renewable and carbon-neutral and will be used as mandatory additive to gasoline in Europe from 2020 (Siegemund et al. 2017). Remarkably, large carriers such as United Airlines have set their goal to operate their entire fleet of airplanes using renewable fuel.

4. Africa's Energy Transition for Off-grid Locations

The dominant fraction of the more than 1.2 billion people on the African continent lives without access to an electricity grid. This is impressively seen in a satellite view of our planet at night where Africa is in pronounced contrast not only with the United States and Europe but also with India (Figure 11.2). No lights are seen on a night-time flight from Johannesburg to central Europe between Gabarone in Botswana and Algeria. Africa is dark at night, with the exception of its northern rim and the populated area in South Africa.

The absence of access to electricity has severe consequences for the living standard:

- The jobs are where the light is on. Access to electricity will greatly reduce the migratory pressure to areas with perceived better living conditions.
- Health depends to a significant extent on access to potable water. Purification, desalination, and recycling of water need energy. Israel, the world leader in the reuse of water, recycles 80% of its household wastewater and uses it mostly for irrigation purposes (Marin et al. 2017). Desalination plants for brackish water and seawater can operate intermittently when energy is available. They are thus an ideal partner of renewable energy.
- Most refrigerators run on electrical energy. Refrigeration is key to keep food fresh and medical drugs safe. In a continent where the average temperatures are high, this is of particular importance.
- Light in a home in the evening is important for family life and learning conditions for students. The ability to connect to the rest of the world by radio, TV, and cell phone is almost synonymous with a high quality of life.
- The African continent represents a very rich cultural diversity (Figure 11.3). Migration from rural areas to big cities or other countries will inevitably lead to its loss. Improved living conditions will allow people to remain in their traditional environment and benefit from their family and tribal contacts. It will protect and conserve their culture, language, and religion. People need a realistic choice about where and how they want to live. Access to electricity allows them to build their own future without sacrificing their identity.

The worldwide commitment to abandon the use of fossil fuels leads to decentralised production of renewable energy. Solar concentrators delivering heat for cooking and hot water generation, as well as solar photovoltaic panels and wind turbines for electricity, can be installed locally. However, to compensate for the cyclic availability of these resources, it is recommended to amend these installations by battery storage. The full cost of solar and wind electricity (not including storage) has dropped below the price of electricity from the grid so that this is also the best economic option. Solar

Figure 11.2 Satellite view of the Earth at night
Source: (colour online, Open source, NASA/NOAA 2009)

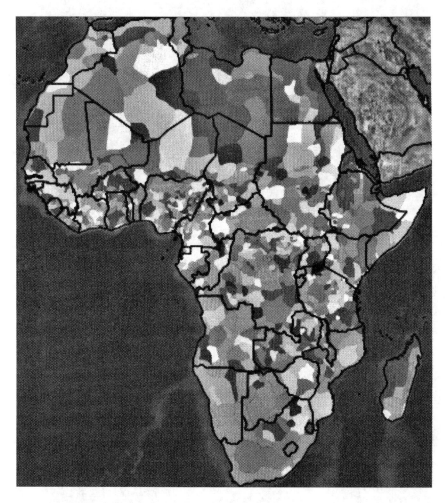

Figure 11.3 Africa's rich and precious ethnic diversity, overlaid with country borders.
Source: National Geographic 2015, open blog, colour online.

panels and concentrators as well as small wind turbines require relatively few skills for installation and maintenance. Nevertheless, engagement and advice by communities and governmental bodies, backed by funding from international institutions is necessary to make this transition.

An interesting and successful solution to manage this energy transition in rural areas has been developed based on private initiative by Bunker Roy in India who founded the *Barefoot College* (Bunker, barefoot), (Bunker 2011). This institution provides practical training to mostly illiterate women and prepares them to install and maintain solar panels and solar concentrators back in their home village. Africa could benefit from such an initiative to provide regional training and funding.

5. Conclusion

Access to energy, in particular to electricity, is a fundamental pillar of today's living standard, but its availability from clean sources represents one of the central challenges to our society. Over the past two centuries, its supply was based on fossil fuel, but the detrimental consequences of this strategy on our climate have now become clear. It is meanwhile widely accepted that we have to stop using fossil fuels before we run out of them, and our reserves, in particular those of coal, should remain where they are. At the same time, the use of wind and solar energy has matured, and the price of renewable electricity has become competitive with fossil fuel-based power. The transition to renewable energy has therefore also become economically viable.

Off-grid locations will skip the stage of extensive use of fossil fuels and enter the age of renewable energy directly. Wind and solar energy are ideally suited for local solutions without having to build an expensive electricity grid. Their cyclic availability may require some transient electricity storage using batteries. However, access to electricity will be essential for human health, well-being, and the preservation of the rich cultural diversity across the African continent.

Note

1 Presented keynote at the workshop, *The Poetics and Politics of Extraction and the Environment*, held at the Centre for the Advancement of Scholarship, University of Pretoria, 28–31 May 2019.

References

Brook, E. 2008. Palaeoclimate: Windows on the greenhouse. *Nature* 453:291–292.
Bunker, Barefoot College International. *www.barefootcollege.org/solution/solar/* (accessed November 17, 2021).
Bunker, R. 2011. Learning from a barefoot movement. *www.ted.com/talks/bunker_roy* (accessed November 17, 2021).
Earth at Night, NASA/NOAA. 2019. *www.nasa.gov/topics/earth/earthday/gall_earth_night.html* (accessed November 17, 2021).
Editorial: Coal in a hole. 2019. *Nature Energy* 4:429.
IRP. 2019. South Africa's integrated resource plan. *www.iea.org/policies/6502-integrated-resource-plan-2019-irp-2019* (accessed November 17, 2021).
Käthelhön, A., R. Meys, S. Deutz, S. Suh and A. Bardow. 2015. Climate change mitigation potential of carbon capture and utilization in the chemical industry, *PNAS* 116:11187–11194.
Knorr, K., B. Zimmermann, S. Bofinger et al. 2016. Wind and solar PV resource aggregation study for South Africa. *www.csir.co.za/sites/default/files/Documents/*

Wind%20and%20Solar%20PV%20Resource%20Aggregation%20Study%20 for%20South%20Africa_Final%20report.pdf (accessed November 17, 2021).

Lüthi, D., M. Le Floch, B. Bereiter et al. 2008. High-resolution carbon dioxide concentration record 650,000—800,000 years before present. *Nature* 453:379–382.

Marin, P., S., J. Yeres and K. Ringskog. 2017. Water management in Israel. *https:// documents1.worldbank.org/curated/en/657531504204943236/pdf/Water-management-in-Israel-key-innovations-and-lessons-learned-for-water-scarce-countries.pdf* (accessed November 17, 2021).

National Geographic, Education Blog, Africa's Dazzling Diversity. 2015. *https:// blog.education.nationalgeographic.org/2015/02/18/africas-dazzling-diversity/* (accessed November 17, 2021).

Paris Agreement. 2015. United Nations climate change: The Paris Agreement. *https:// unfccc.int/process-and-meetings/the-paris-agreement/d2hhdC1pcy* (accessed November 11, 2021).

Siegemund, S., P. Schmidt, M. Trommler et al. 2017. *The Potential of Electricity-based Fuels for Low-emission Transport in the EU*. Berlin: Deutsche Energie-Agentur GmbH (dena). *www.dena.de/fileadmin/dena/Dokumente/Pdf/9219_E-FUELS-STUDY_The_potential_of_electricity_based_fuels_for_low_emission_transport_ in_the_EU.pdf* (accessed November 17, 2021).

Solar GIS. 2011. DNI map South Africa. *www.researchgate.net/figure/13-DNI-map-Africa-a-South-Africa-b-Solar-GIS-2011_fig14_299604949* (accessed November 17, 2019).

Utembe, W., E. M. Faustmann, P. Matatiele and M. Gulumian. 2015. Hazards identified and the need for health risk assessment in the South African mining industry. *Human and Experimental Toxicology* 34:1212–1221.

Winkler, H. 2018. Reducing inequality and carbon emissions: Innovation of developmental pathways. *South African Journal of Science* 114:1–7.

12 Poetics and Politics of Resource Exploitation in Africa

Insights From Chapters

James Ogude and Tafadzwa Mushonga

12.1. Introduction

Within the broad framework of environmental humanities, the previous chapters have integrated perspectives from history, philosophy, political ecology, political geography, literature, science, and law to unpack the nature of Africa's environmental struggles. In the introduction, we argued that the dispossessing nature of resource exploitation is at the core of what defines the perverse nature of Africa's environmental struggles. We went on to identify these struggles as revolving around four thematic thrusts; environmental (in)justice, violent capitalocenes, repression of indigenous knowledge, and climate change, which in this conclusion we argue are Africa's biggest predicament. We now distil some of the key issues raised throughout the volume to reiterate the politics of resource exploitation in Africa and how these define the current and future environmental status.

12.2. The Persistent History of Resource Exploitation in Africa

The four thematic thrusts identified as underpinning resource exploitation in Africa are largely traced back to the legacy of colonialism in Africa, which was defined by environmental racism, disenfranchisement, and dispossession at its ideological core. The histories of exploitation presented in this volume demonstrate the temporal continuities of exploitation and struggle, dating back from the advent of colonialism to the present neo-colonial moment. Such histories are critical for the manner in which they help us to understand current environmental predicaments on the continent as predominantly an issue of the Humanities. The evidence gleaned from chapters is also a sharp reminder that Africa's resources and their exploitation continue to rest in the hands of external actors – a legacy of colonialism. The same evidence also suggests that, unless radical changes take place in relation to state policies and legal frameworks, and recognition of traditional practices that have withstood the test of time, the continent will continue to fall victim of resource exploitation, woefully, to the detriment of its citizens. Just as

DOI: 10.4324/9781003287933-12

today is informed by the past, today becomes inherently the future's history. Following this pattern of change and without any deliberate disruptions to history, it is likely that histories of resource exploitation and environmental struggles will continue unabated. It is for reason that the contributions in this volume argue that unless the cycle of exploitation, dispossession, and disenfranchisement are broken, Africa's environmental woes will persist into the foreseeable future. How then do we break this cycle in order to bring about the desired change and reverse what may look like Africa's fate?

Although the chapters in this volume make no claim to offer any solutions to these difficult questions, we believe that in drawing attention to the protracted history of resource exploitation in Africa and the resultant environmental degradation that comes with it, they point to how our environment could be treated differently. Significantly, they link the problem of resource exploitation to the history of the empire (see, e.g., Larteza and Sharp 2017), while equally showing how we are implicated through post-colonial state complicity. Several chapters have highlighted the general lack of commitment by African state actors to address the issue of resource exploitation and related environmental and social issues the continent faces. If there are any lessons to be drawn from the impact of the Covid-19 pandemic it is that our resilience systems have been weakened over the years and a stark reminder of state complacency over many years in addressing the social inequalities brought chiefly by the continent's resource exploitation. The evidence presented, in fact, demonstrates how resource exploitation on the continent has transitioned from exploitation by coercion to exploitation by consent, as McKay (2017) would have it, and consequently, how such exploitation becomes a state-led and state-sanctioned activity. The state in Africa sanctions exploitation of national resources in the name of development, while undermining the very principles of development it purports to achieve; firstly, by treating its people as subjects rather than citizens, and secondly, by not holding corporate and multi-national offenders of environmental and social degradation to account.

Further to these, what emerges from chapters is a general awareness of environmental issues, aspirations for change, and above all that we can marshal, at least from our recent past, the will to promote the principles of co-responsibility, earth-keeping, and conviviality that have been the cornerstone of many African communities for generations. But while this general awareness and public debate on such issues is a key element of broader progressive environmental change, it is not getting the support it needs from the state. Outa, Masolo, and Roduner, in this volume, argue that the state should lead the struggle to save Africa from exploitation and/or its adverse environmental effects. Following their arguments, we reiterate the point that has been made by other scholars (see Steyn 2002), that by not taking the lead in saving Africa from resource exploitation and its myriad environmental struggles, state complacency will in itself become an environmental problem that the continent will have to grapple with for many years to come.

Another key insight linked to resource exploitation and one that cuts across all the four themes explored in this volume is resource sovereignty. It can be discerned from the chapters that while African states have national–territorial control over resources on the continent, the privilege to exercise control has not been extended to citizens and local populations as Crawford and Botchwey (2016), have also noted. The issue of land, land tenure and resource appropriation touched on by a number of chapters expose people's general lack of sovereignty over resources. While citizens watch resources being taken away from them by the political elite, they remain impoverished and often forced to resign as the plunder of their resources continues unabated. This, as we have pointed out, is in part a legacy problem linked to colonial statecraft and patterns of exploitation, but it is also a problem of the neo-colonial state in Africa, which continue to sanction, what Nixon (2007) calls slow violence, against its people and their environment.

Contrary to the assertion by Emel et al. (2011), assertion, we argue that resource sovereignty cannot be realised outside the armpits of the state, while acknowledging that non-state actors, also have huge control over resources. But as we have also discussed previously, the state in Africa is hugely implicated when it comes to ceding control over resources to external interests, especially where capital is involved. We opened this volume with the dreadful death of a community member who died in defence of their environment from exploitation. Lack of resource sovereignty and state support arguably led to her death, and we think such incidents are a representation of the continent's predicament and the absence of autonomy when it comes to resource control. We argue that the continued lack of resource sovereignty amongst the African population, and the unremitting hegemonic role of the state in the governance of resources (and people), are some of the most enduring issues on the continent. This is precisely because the lack of control over resources by the African people is likely to exacerbate problems of climate change adaptation and the preservation of indigenous environmental practices. It is further likely to compound social and environmental injustices and the concomitant eruption of resource-based violence as all chapters in this volume have endeavoured to highlight.

12.3. Repressed Knowledges and African Environmental Problems

Alongside the erosion of indigenous knowledge and practices, the perpetuation of violent capitalist models, enduring environmental (in)justice, and climate change that continue to take a toll on the environment and human economies, the chapters in this book also suggest that Africa's bleak future, and its resource extraction history, could be turned around by drawing on local practices, often ignored in global environmental discourses. What is evident is that top-down approaches, especially those coming from the Global North, will not on their own resolve climatic problems in Africa. Our volume points to three important trajectories in this respect. The first pertains to change in state ideology about people and resources. One of the issues that stand out is that while resources bring much-vaunted economic growth, development in Africa cannot be sustainable if exploitation and degradation of the environment continue unchecked. Any development that fails to consider people's welfare and ways of being in the world, is likely to fail. Development cannot be realised by overexploiting and damaging the environment. Given the wide-ranging connection between resource exploitation, environmental degradation, environmental injustice, the state in Africa must respect and recognise the rights of its people in relation to their environments, traditions, knowledge systems, and needs, without which sustainable development is impossible. From an environmental humanities perspective, development can rather be achieved if entrenched in multi-species justice, that is, justice for both human and non-human bodies.

The second issue is that the African state must not rely on imported modes of development, most of which may not be attentive to local nuances. They have to consider the significance of African traditions and cultures as part of the key driving scientific and sustainable governance approaches to resource-linked development (Ikuenobe 2014). In this regard, the state should encourage the synergy between local and global knowledge systems, to create more sustainable and context-based solutions for Africa as Guluma and Dantamo argue in Chapter 8. Such recognition must also really begin with recognising African societies' indigenous knowledge and culture as part of the national legal frameworks as Outa in Chapter 10 points out. Besides, subjugated African concepts such as *Ubuntu* may well offer us what Achille Mbembe calls "an enquiry into the possibility of a politics of the future, of mutuality and of the common" (2009: 35), ways of framing our conviviality and interdependence in the modern world. As Ogude and Masolo in Chapters 6 and 9, respectively argue, these changes need ethical and moral leadership rooted in the experiences and rich histories that African societies continue to offer.

As we have highlighted in the introduction, the problem of climate change is one of the key challenges facing Africa. Contributions in this volume also touch on the climate change crisis and suggest the kind of solutions the continent needs in this regard. The recent COP 26 in Glasgow resolved to "phasing down" rather than "phase out" the use of fossil fuels such as coal. Critics think that the change of phrasing to phasing down is a green light for ongoing coal development (Rusell 2021). But what would this mean for Africa which is still largely dependent on fossil fuels for energy? There are increasing calls for African governments to accelerate the energy transition. And yet, as we well know, the transition for Africa to clean energy requires more than just the political will, but also the vision and resources to balance this with our economic capacity to make such changes without exposing our vulnerabilities. Political will is important in reducing the state's reliance on international institutions for funding, but on its own, it may not be sufficient if it is not balanced with a clear programme of economic sustainability rooted in local needs and a prudent and responsible use of Africa's resources.

12.4. Conclusion

Finally, we need to remind our readers that one of the many goals of this book is to surface a range of perspectives on resource exploitation and to foreground the continued, struggles in post-colonial Africa. As demonstrated throughout the volume these struggles manifest in the manner in which Africa continues to succumb to violent capitalocenes, vulnerability to climate change, erosion of indigenous knowledge practices, and environmental injustices that ensue. An overall insight emerging from this volume is that the vulnerability of African countries will remain a concern if the colonial architecture tying down the continent to exploitative ethos and western ideas of modernity are not redesigned to suit the needs of the African continent. The importance of leadership reverberates throughout the book, with the hope that African leaders will facilitate changes in the best interest of the continent's environments, resources, and citizens. However, given the uneven geopolitical relations between Africa and the Global North, in particular, Africa's historical role as the supplier of raw materials to industrialised nations of the West, a major mindset is needed to re-centre Africa's environmental priorities and to end the ruthless degradation and exploitation of Africa's resources and its people. This is a challenge that Africa's vast communities must undertake and re-assert their agency as the custodians of their resources and environmental wellbeing.

References

Crawford, G. and G. Botchwey. 2016. Foreign involvement in small-scale gold mining in Ghana and its impact on resource fairness. In *Fairness and Justice in Natural Resource Politics*, ed. M. Picher, C. Staritz, K. Küblböck, C. Plank, W. Raza and F. Ruize Peyré, 193–211. London and New York: Routledge.

Emel, J. M., T. Huber and M. Makene. 2011. Extracting sovereignty: Capital, territory, and gold mining in Tanzania. *Political Geography* 30:70–79.

Ikuenobe, P. A. 2014. Traditional African environmental ethics and colonial legacy. *International Journal of Philosophy and Theology* 2:1–21.

Larteza, V. and J. Sharp. 2017. Extraction and beyond: People's economic responses to restructuring in southern and central Africa. *Review of African Political Economy* 44:173–188.

Mbembe, A. 2009. Postcolonial thought explained to the French: An interview with Achille Mbembe. *The Johannesburg Salon 1. www.jwttc.org.za /the_salon/volume_1/achille_mbeme.htm.*

McKay, B. M. 2017. Agrarian extractivism in Bolivia. *World Development* 97:199–211.

Nixon, R. 2007. Slow violence, gender, and the environmentalism of the poor. *Journal of Commonwealth and Postcolonial Studies* 13:3–12.

Rusell, C. 2021. Coal trajectory is set whether its 'phase out' or 'phase down': Rusell. *www.reuters.com/business/cop/coal-trajectory-is-set-whether-its-phase-out-or-phase-down-russell-2021-11-14/* (accessed November 2, 2021).

Steyn, P. 2002. Popular environmental struggles in South Africa, 1972–1992. *Historia* 47:125–158.

Index

Note: Page numbers in *italics* indicate a figure and page numbers in **bold** indicate a table on the corresponding page.

AAC *see* Anglo American Corporation (AAC)
Aari of Southern Ethiopia: communities 134; culture and identity 126; ecological thoughts 135; indigenous knowledge 127; knowledge 135; map of 125, *125*; people 125; practices 135; traditional religion 127–128, 130
Acacia bushes 51
Achebe, C. 153–154, 163–165
Adorno, T. 98
aesthetics of proximity 97
Africa/African 2; arid land in 11; British colonialism in 105; climate change 13, 181–182; CO_2 emissions 148; colonial marginalisation of 14; communities in 2–3, 10–11; context and needs 8, 16; cosmology 164–166; ecology 123; economic and political developments in 76; economic history of 117; environment 2–3, 14, 16, 117, 124, 179; indigenous religions in 172; land tenure systems in 117; mining regions of 15; natural resources in 3, 12; palm oil extraction in 107; pastoral regions of 139; pervasive problem in 13; resource exploitation in 7, 16; traditional societies in 15
African goats 44; commercial slaughter of 49; impound notes of stray 43
Africa Partnership Forum 11
Agricultural Marketing Authority of Southern Rhodesia 49

agro-ecological zones 134–135
Algeria, pastoralist economy of 41, 147, 174
Anglo American Corporation (AAC) 76, 79
Angola 11, 27; Provinces, chronic undernutrition by 29, 30; San communities in 28
Angora goats 44–45
Antarctic ice core analysis 170
Anthropocene 3, 106–108, 117; concept of 26–27; contemporary environmental crises of 107; debate 6; geographies of 108; label masks, effects of 107
anthropocentrism 24, 128; critique of 14, 23; ecological crisis 6; historical celebration of 25
anti-goats: conservation 39; environmentalism 40; zeitgeist 38
anti-poaching operations 67
artificial nature/culture dichotomy 2
Ashti forest 132
atmospheric warming 145
attritional violence 107
authoritarianism 7, 13, 108
Ayelazuno, J. A. 3

Bantu community 28
Basel Convention of 1989 82
BEAs *see* Bilateral Environment Agreements (BEAs)
Beatty, A. C. 76
Berlin Conference 27, 115
Berry, S. 117

Biden, J. 162
Bilateral Environment Agreements (BEAs) 157
biodiversity 56–58, 65, 70, 127
Black Lives Matter movement 140
Boer goats 44–45
Botchwey, G. 181
Botswana 30, 110
Bowles, Chester 162
British South Africa Company (BSAC) 59, 76
Brockington, D. 58
Brooks, J. S. 130
Brundtland Commission of 1984 82
BSAC *see* British South Africa Company (BSAC)
Butler, J. 26, 31, 34

C-19 global pandemic 35
Cacimba 31, *31*
Cape of Good Hope forestry law 42
capitalism 7, 9, 100
capitalist democracy 100
capitalocenes: conceptualisation of 7; expansion of 12; principal drivers of 8; theoretical perspectives on 7
capriphobia 38
carbon dioxide emission 13, 172–173
Cela settler project 28
Centre for Environmental Rights 1
Cernea, M. M. 10
Cession, treaty of 115
Chakrabarty, D. 103
Chansa, C. J. 14–15
China 16, 149–150
civil conflicts 7
climate change 12, 145–146, 165, 181; in Africa 13; effects of 107; geographical distribution of 12; mitigation 5
Climate, Community and Biodiversity Alliance (CCBA) standards 156–157
climate injustice 139–151
climate justice 52; critical questions of 5; movements 13
coal mining 1–2, 173
colonialism/colonial 2, 5, 102, 117, 123–124; in Africa 179; domination 105, 111–112; incursion 108; Kenyan societies by 95; land tenure systems 118; legacy of 179–180; modernity 99–101; power, economic development of 30; restrictions of 93; school systems 113; territorialisation 5; Western 164
colonialist ideology 106
coloniality 7, 14
commercial farming 135
commercial poachers 58–59, 63, 65, 67
contemporary environmentalism 3
contested ecologies 6, 9
Copperbelt mines 75–79, 81–82, 85, 87; environmental management on 84, 87; excessive pollution 88; mining pollution 87; sale of 84; Zambia 14
Corporate Social Responsibility (CSR) 88
cosmological violence 108
Cradle of Humankind *171*, 172
Crawford, G. 181
creation, biblical story of 172
creativity/creative: legal solutions 153, 156; literatures 154, 163–164; solutions and approaches 157
criminalisation 64
Crosby, A. 97, 124
Crowther, S. A. 112–113
Crutzen, P. 25
CSR *see* Corporate Social Responsibility (CSR)
cultural violence 68–69
Culture and Imperialism (Said) 93–94

Dantamo, Z. J. 182
de Chardin, T. 23, 34
decision-making process 4–5, 135
decolonisation, process of 124
deforestation 8, 38–39, 99–100, 111, 131, 165
DeLoughrey, E. 26, 93–94, 103
de Matos, N. 27
developing countries 32–33, 62, 84, 157–158, 162
development projects 3, 10, 111, 117
Dialectics of Enlightenment (Adorno) 98
discrimination 29, 39, 151
disenfranchisement 14, 179–180
dispossession 2–3, 5, 7, 14, 51, 58, 179
distributive justice 4
Doro, E. 14
drought 15, 29, 95, 101, 103, 126, 134, 139–140, 144–145, 149–150
Duffy, R. 69
Dutta, S. 9

186 Index

Earth, satellite view of 174, *175*
Earth Charter 1992 23–24
ecocide 158–159
eco-critical awareness 163
ecocriticism 26, 93, 97
ecology/ecological 102; balance 99; boundaries 147; crisis 6, 10–11; degradation 94; disaster 93, 95; imperialism 97, 124–126; knowledge 123
economic deprivation 13
economic exploitation 7–8
electricity 170, 174–175
Emel, J. M. 181
endangered species 161–162
energy projects 4–5
energy transition 170; carbon dioxide emission 172–173; North-eastern South Africa 170–172; off-grid locations 174–177; in rural areas 176
"enset" farms 126–127
Environmental Council of Zambia 82
environment/environmental: balance 108; catastrophe 165; change 180; conservation and preservation of 154; control 143–151; costs, distribution of 4; crisis in Africa 16; degradation 5, 15, 182; exploitation 3; globalisation 93; humanities 3, 7, 27, 179, 182; injustice 181–182; legislation 81; management 76, 85; policies, formulation of 85; pollution, conditions for 75; protection 12, 153, 155; racism 14, 39, 179; struggles, histories of 180; sustainability 102; violence 107–108
Environmental Humanities of Extraction (EHfE) 3
environmentalism 93, 102–103; centrality of 97; ecology and 94
environmental justice 2–6, 14, 23–24, 39; advocacy 14; movements 6, 9; struggles for 5
Environmental Justice Organisation 1–2
Environmental Protection and Pollution Control Act (EPPCA) 82
epistemic injustice 150
epistemic violence 108
EPPCA *see* Environmental Protection and Pollution Control Act (EPPCA)
ethnic diversity *176*
ethnic groups 126
ethnic identity, formation of 105

European Enlightenment 103
Europeanisation 123
Eve, T. 108
exclusionary resource extraction 68
Eyong, C. T. 10

Fairhead, J. 3, 106
famine 139–141, 144–146
Fanon, F. 111–112
feudalism, indoctrinations of 126
flash floods 146
food security 8, 32, 111, 143
forced displacement 33
forest: dwellers 161; ecosystems 166; legal framing of 57; protection 153; quotas 165; regeneration 45; sacred sanctity of 164–166; sanctity of 165; tragedy and status of declining 155–156
Forest Protection Unit (FPU) 65
forestry governance regime 157
forestry preservation 41
fossil fuels 170, 174, 177, 182
Foucault, M. 25, 98, 111–112
FPU *see* Forest Protection Unit (FPU)

Galtung, J. 7, 68–69
geographical identity 124
geographical violence 123–124
Gikandi, S. 94
global capitalism 26, 32, 34, 123–124
global commons 158–159
Global Health Security Index (GHS Index) 35
globalisation 102, 109, 123–124
global North 16, 146, 148, 181, 183
global survival 150
global warming 13, 147–148, 165
goats 38–40; discriminatory taxation against 41–42; in livestock hierarchy 40; polemicisation of 40
Godlonton Report 48
Goldie, G. T. 115
Goldman, M. 42
Gomes, C. 14, 16
Gray, A. 75
Green Climate Fund 162
green colonialism 57–58
green credentials 3
green house gas (GHG) emissions 148, 150, 155, 157, 162
Guluma, H. 15, 182
Gyekye, K. 123

Hallen, B. 144
Hallowes, D. 5
Handley, G. B. 93–94
Harvey, D. 3
Hayford, C. 116
hazardous substances 78, 80, 82
Hecht, G. 107
Holm, P. 9
homegrown calendar 134–135
Horkemer, M. 98
Horn of Africa 13, 145
human–environment relationships 115
human exceptionalism 14
humanism, ideology of 79
humanities, apt relevance of 153
human–nature dichotomy 15
human needs 150, 164
human rights: of natives 142; violations 69
Human Rights Watch 1
hunger 32, 103, 139–141, 144–145
Hwange national park (Hwange) 59
hydrogen, electrolytic production of 173

IEL *see* International Environmental Law (IEL)
Igoe, J. 58
Iheka, C. 97
Ikuenobe, P. A. 124
illegal resource extraction 65
imperialism, ideologies of 125
indigenous communities 2, 15–16
indigenous cosmology 15
indigenous groups 105
indigenous ideology 110
indigenous knowledge 8–11, 13, 127, 135, 143, 153; forms of 114; resources 9; sustainability of 10; systems 10
indigenous peoples 9, 161
indigenous religions 172
indigenous scholars 108–109
indigenous science, wisdom and creativity of 106
Indonesia 147, 149, 155, 162
industrial agriculture 29, 31
industrial growth 24, 147
industrialisation, commodity and resource for 8
industrial knowledge 150
Integrated Resource Plan 173
International Court of Justice (ICJ) 156
international criminal tribunals 158

international environment agreements 156
International Environmental Law (IEL) 154, 156, 163; Article 2 of Forest Instrument 160; creative legal solutions 153; creative literatures 163–164; creative solutions and approaches 157; endangered species 161–162; forestry protection approaches 156–157; indigenous peoples, forest dwellers, and local communities 161; international environment protection regime 154–155; re-conceptualising "common but differentiated responsibilities 162; redrafting Article 5 160–161; sacred sanctity of forests in African cosmology 164–166; state sovereignty and expanding international jurisdiction 158–159; strengthening role of non-state actors 163; structure and outline 153–154; tragedy and status of declining forests 155–156
international forest protection regimes 164–165
International Timber Organization (ITTO) 157, 163
International Tropical Timber Agreement (ITTA) 156, 163
invented nation 112
ITTA *see* International Tropical Timber Agreement (ITTA)
ITTO *see* International Timber Organization (ITTO)
ivory poaching 57, 63, 69–70

Jackson, G. C. A. 75
Jaganyi 141, 143–144, 150; healing 145; individual evidentiary criteria for 144; rain-making powers 145
Jagre, Z. 15
Johnson, S. 109
jurisdictional zones 159, 166

Kavango Zambezi Transfrontier Conservation Area project (KAZA-TFCA) 62, 65
Kenya 98–99, 123, 144–147
Kidane, B. 126
killing of trees 131
knowledge: indigenous forms of 9; individual evidentiary criteria for 144

188 Index

Korieh, C. 118
Kwashirai, V. 57

land 15, 110; in Africa 108; alienation 95; capitalist appropriation of 33; centralisation and conservation 39; commercialisation 111; grabbing 32; ontological value of 96; ownership 115; spiritual significance of 96; surplus economy 117; tenure 181; usurpation and expropriation 28
Land Apportionment Act of 1930 46
landlessness 3, 6, 10
Leach, M. 106
Legesse, B. 9
Leverhulme, W. 118
Lever, W. 113–114
literary ecocriticism 26
Livingston, J. 110
local communities 8–9, 28, 32, 48, 61, 63–64, 69, 161
Lunstrum, E. 58
Lynn, M. 107

Mamdani, M. 117
Manichean categorisation 157
Mariam, M. H. 146
Martinez-Alier, J. 93
Masolo, D. A. 182
Masse, F. 58
Mbembe, A. 182
McKay, B. M. 180
MEAs *see* Multi-Lateral Environment Agreements (MEAs)
medicinal plants 126–127, 130
Merensky reef 172
Mfolozi Community Environmental Justice Organisation 1–2, 4–6, 8
mining industry: air pollution 81; on environment 77–78, 82; infrastructure and equipment 80; laws in Northern Rhodesia 78; ownership structure of 79–80; privatisation 84; techniques and disposal methods 79
Moloney, A. 105–106, 113, 116
Moore, J. 6–7
Mount Kenya 146
Mount Kilimanjaro 140, 146–147
Movement for Multiparty Democracy (MMD) government 82, 84–85
Mozambique 27, 32
Mufulira Mine in 1970 80

Multi-Lateral Environment Agreements (MEAs) 157
Munnick, V. 5
Mushonga, T. 14–15

Naess, A. 23
national sovereignty 58
Native Land Husbandry Act (NLHA) 46, 49
natural disasters 33, 133
natural environments 97, 128, 130, 136
natural resources 3, 8–9, 28, 32–33, 81–83, 85, 100, 106, 108
Ndebele, N. 7
net-zero carbon emission economy 170
Nguni group of languages 151
Nigeria: "British occupation" of 115–116; conflicts 117; Land and Native Rights Ordinance 118; Land Use Decree 118; market demands and colonial policies 114; mining and extraction of oil and minerals 108; palm oil exploitation in 111; Yoruba identities in 9
Nigerian Oil Palm Produce Marketing Board (NOPPMB) 118
Nixon, R. 7, 100, 107, 181
Noguchi, M. 128, 135
Non Legally Binding Agreement (NLBA) 154, 160
non-timber forest products (NTFPs) 59
Northern Rhodesia 78; copper production in 76; open-pit mines in 77
NTFPs *see* non-timber forest products (NTFPs)
Ntshangase, F. 1
nuclear wastes 148
nuclear weapons 107

Ogude, J. 15, 151, 182
Oluwole, S. 109
Ongode, A. 142–143
ontological exceptionality 25
ontological violence 108
OPEC, formation of 147–148
Oruka, H. O. 123
"Osembo" hunger and famine 144
Outa, G. 16
outmigration 33
Oyewunmi, O. 114
Oyo kingdom 109

Index

Painted Dog Conservation 67
palm oil commercialisation 111
Palsson, G. 26
Paris Agreement on Climate Change 148, 151, 154, 159–160, 170
Parker, R. J. 75
pasture management 50–51
Peace Parks Foundation (PPF) 62, 65
Peluso, N. L. 57
phenomenology 26
plant biodiversity 126–127
plantation: economies 6; staggering of 166
platinum-group metals 172
pledged reforestation 165
ploughing "virgin forests" 164
Plumwood, V. 124
poaching crisis 57–58, 65
political commitment 162
political ecology 3, 7, 26, 29, 36, 56, 179
political will 182
post-colonial ecologies 93–103, 124
post-colonialism 26, 95
poverty 99, 164
PPF *see* Peace Parks Foundation (PPF)
predatory capitalism 32
private safari operators 67
procedural justice 4
proliferation of violent capitalocenes 6–8
protected areas, exclusion and violence in 56–57; converging interests 59–64; divergent values, politics of exclusion and 64–68; Sikumi Forest 59; territorialisation, securitisation, and militarisation 57–58; violent exclusion and implications for conservation 68–70

racism 13–14, 119, 124, 151, 179
rainfed agriculture 12
rationalisation, violent process of 98
regional training 176–177
religious institutions 135
renewable energy 170, 173–174, 177
renewable fuel 173
resource appropriation 2, 181
resource exploitation 2, 4–5, 7, 11, 13–14, 179–180, 182; consequence of 10; history of 179–181; issue of 180; projects 3; repressed knowledges and environmental problems 181–182; violent modes of 7

resource extraction: history 181; practices 68; structural and physical barriers in 66
resource scarcity 7
resource sovereignty 181
resource utilisation 14
restorative justice 4
Rhodesian Selection Trust (RST) 76, 79
Rhodes Must Fall movement 140
Roduner, E. 16
Ross, M. L. 75

Sack, Robert 115
Said, E. 93–94, 123–124
Samuel Johnson and Samuel Ajayi Crowther 112–113
San community 28, 30
Savory, E. 163
scapegoat 40
Schumaker, L. 75
scientific knowledge 143
seasons 134–135
Segui, E. 38
sharecropping 165
Shigeta, M. 128, 135
Siddle, D. 40
Sikumi Forest 58, 63, 67, 69; biodiversity in 59; commercial poachers 63; communities surrounding 63–64; for conservation 63; converging actors and interests in 59, 60; crop production 64; livestock production 64; non-governmental organisations 62–63; private safari operators in 61–62; resources in 59
silicosis 76
Smallpox 144–145
Smoke Damage (Prohibition) Act of 1935 78
socialism 151
social justice 15
social stratification 13
social structures, destabilisation of 9
social violence 7
Sodipo, J. O. 144
Soil Conservation Committee 41
solar energy 170
solar photovoltaic energy 173
South Africa 1, 11, 13, 30, 151; energy transition 172–173; ferrochrome industry 1–2; mining map 170, *171*
Southern Ethiopia 123–124; Aari people 125–127, 135; administrative

structures 125–126; analysis and discussion of results 127–128; green leaves and shades 133–134; homegrown calendar, seasons, and agro-ecological zones 134–135; huge trees 130–131; respect/power, holiness, and nature 132–133; sanctified groves, forests, springs, and places of origin 128–130; *S'oyesi* forests 131

Southern Rhodesia 42–46; African goats in 42–46, 49, **50**; bush encroachment 50; Cold Storage Commission (CSC) 49; colonial conservation and goat question in 46–49; conservationists in 51; ecological validity 50; environmental landscapes in 49–51; goat farmers 45–46; goat question in 46–47; Land Apportionment Act 46; Native Land Husbandry Act 49; Native Trade and Industrial Commission of 1944 48; statistics of goats in 45, **46**

Southwestern Nigeria 105–108; colonial tools of environmental domination 114–117; environmental violence in 108; indigenous groups of 110; land and land tenure system 117–119; precolonial 116; "Yoruba" human–environment kinship 109–111; "Yoruba" in South-Western Nigeria 111–114

sovereignty 58, 157, 181
Soyinka, W. 123
space science 143–151
Spatial Monitoring and Reporting Tool (SMART) technology 67
state capital, colonial legacies of 7–8
state policy, irrationality of 32
state power, colonial model of 8
state responsibility 159
state sovereignty 157
Steyn, L. 1
Stockholm Declaration 154, 159
Stoemer, E. 25
Sudan 11, 32, 41–42, 126
surplus land 15
survival migration 33
Sustainable Development Goals 33
sustainable forest management 153, 157, 164
systemic racism 13

Tanganyika 51, 54, 140–143
Tendele Coal Mining 1–2
territorialisation 57, 114–115; conceptual aspects of 58; process of 58; treaties and 116–117
TFA *see* Things Fall Apart (TFA)
Things Fall Apart (TFA) 163–166
Thiong'o, Ngugi wa 5, 10, 15, 94–96, 101, 103, 123
Tiffin, H. 124
time 143–151
Tolessa, K. 126
transboundary water resources 157–159
Trump, D. 148, 151
Tutu, D. 151

Ubuntu, principle of 102, 151
Uganda 140–141
undernutrition 32
UN Framework Convention on Climate Change (UNFCCC) 154
UNIPs *see* United National Independence Party (UNIPs)
United National Independence Party (UNIPs) 79–82
United Nations Conference on the Human Environment in 1972 81
United Nations Environment Program 82
universal citizenship rights 29
urban sanitary boards 43

vanadium resources 172
Vandergeest, P. 57
violence: emergence of 64–68; in mining sector 7; spatiality of 14
violent capitalocenes: conceptualisation of 7; expansion of 12; principal drivers of 8
violent exclusion for conservation 68–70
violent resource exploitation 8
virgin forests 164

waste disposal method 77
waste management strategies 77
water resources 11–12
watershed management 61
West African War Council Supply and Production Committee 111
Western colonialism 164
white supremacy 57–58

Williams, P. 99
wind energy 170, 173
Witwatersrand basin 170–171
wool production 40
World Climate Change Conference in Madrid in December 2019 148
World Commission on Environment and Development 157
World Meteorological Organisation (WMO) report 11, 81
wretched animals 42
Wright, L. 99–100

Yang, W. 108
Yntiso, G. 128, 135
Yoruba 105, 108; ethnic identity 112; gender categories for 114; history 112–113; human–environment kinship 109–111; identities in Nigeria 9, 15; indigenous groups 109; land 115, 118; language 112; origin and cosmology 113; political identity 115; in South-Western Nigeria 111–114

Zambia 75–76, 81; Act and Development Agreements 85; Copperbelt 87–88; Disaster Management Act of 2010 85; Environmental Management Act 85; extractive industry 14; health, and safety on 76–79; Mines and Minerals Act 79, 82, 86; Mines and Minerals Development Act of 2008 84–85; mining facilities in 9; mining, pollution, and environmental regulation on 84–86; mining regulation during UNIP rule 79–82; Mining Regulations 1971 and 1973 80–81; MMD rule, privatisation, and environmental management in mining industry 83–84; National Conservation Strategy in 1985 82; natural resources and environmental measures in 82; political economy 14–15; Water Resources Management Act of (Number 21 of 2011) 86; Wildlife Acts of 2017 86
Zambia Consolidated Copper Mines (ZCCM) 82
ZCCM *see* Zambia Consolidated Copper Mines (ZCCM)
Zeh, W. 40–41
Zimbabwe: Environmental Management Agency (EMA) 62; forest reserves in 69; goat farming in 14; Hwange Sanyati Biological Corridor Project (HSBCP) 62; non-governmental organisation (NGOs) 62; protected forests in 14; Sikumi Forest Reserve in 56
ZimParks 59–62

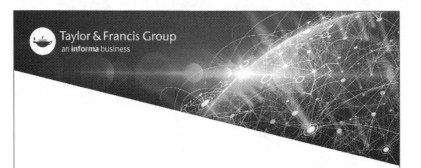